CONSERVATION BY THE PEOPLE

CONSERVATION BY THE PEOPLE

The Future of Biodiversity in a Divided World

TAYLOR DOTSON

The MIT Press
Cambridge, Massachusetts
London, England

The MIT Press
Massachusetts Institute of Technology
77 Massachusetts Avenue
Cambridge, MA 02139
mitpress.mit.edu

The MIT Press would like to thank the anonymous peer reviewers who provided comments on drafts of this book. The generous work of academic experts is essential for establishing the authority and quality of our publications. We acknowledge with gratitude the contributions of these otherwise uncredited readers.

This book was set in Adobe Garamond by Westchester Publishing Services. Printed and bound in the United States of America.

Library of Congress Cataloging-in-Publication Data

Names: Dotson, Taylor author
Title: Conservation by the people : the future of biodiversity in a divided world / Taylor Dotson.
Description: Cambridge, Massachusetts : The MIT Press, [2026] | Includes bibliographical references and index.
Identifiers: LCCN 2025024100 (print) | LCCN 2025024101 (ebook) | ISBN 9780262052092 paperback | ISBN 9780262052108 pdf | ISBN 9780262052115 epub
Subjects: LCSH: Biodiversity conservation—Social aspects | Biodiversity conservation—Political aspects | Biodiversity conservation—Citizen participation
Classification: LCC QH75.A3 D68 2026 (print) | LCC QH75.A3 (ebook)
LC record available at https://lccn.loc.gov/2025024100
LC ebook record available at https://lccn.loc.gov/2025024101

EU Authorised Representative: Easy Access System Europe, Mustamäe tee 50, 10621 Tallinn, Estonia | Email: gpsr.requests@easproject.com

Contents

1 WILD POLITICS

Between the cacti and on a nearby highway overpass, right-wing militia snipers lie in wait. Their rifles' crosshairs land on Bureau of Land Management (BLM) officers and National Park Service rangers. Sheltering behind a row of SUVs, the feds have donned tactical gear and carry AR-15s. The government has its own snipers in place, as court documents will later reveal. They are prepared to end the lives of American citizens, should the confrontation escalate. Southwest of the aptly named town of Bunkerville, officers surround local rancher Cliven Bundy's home with armored vehicles.[1]

What sparked the standoff was a simple tortoise, *Gopherus agassizii* to be exact. The chihuahua-sized reptile spends most of its life in a dormant, sleep-like state, crawling out of its den during the warmer and wetter months to munch on grasses, wildflowers, and cacti (figure 1.1).[2] The species was listed as threatened with extinction in 1990. Afterward, the local government, the BLM, and The Nature Conservancy (TNC) worked together to buy up local grazing permits. Complying with the Endangered Species Act (ESA), these groups agreed, would mean far fewer cattle on the federally owned lands that ranchers paid to access. Cliven Bundy was the last holdout. When the conditions of his permit were modified by the BLM, he stopped paying grazing fees altogether.[3]

After years of courthouse battles, the BLM made the area around Bunkerville a "cattle trespass closure area." The agency began impounding Bundy's herd and demanded back payment of more than $1 million. Bundy holed up at home and decried governmental tyranny while his family and

Figure 1.1
A Mojave desert tortoise inspects the fruit of a prickly pear cactus, blissfully unaware of the human conflicts raging all around him.

supporters openly clashed with federal officers. Footage of his son, Ammon, being tasered spread on social media. Hundreds of militiamen and other volunteers poured in from around the nation. As one participant later recounted, "Fingers [were] on triggers on both sides. If a car had backfired, the shooting would have started."[4]

Citing public safety but perhaps fearing a repeat of the tragic violence of Waco, Texas, and Ruby Ridge, Idaho, federal officers backed down and released the impounded cattle. In fact, Bundy continues to flout BLM decisions and graze his herd in Gold Butte National Monument to this day. Here, and elsewhere, the conflict between using and preserving the landscape is at a stalemate.[5]

This book isn't about Cliven Bundy, at least not directly. It deals with the scientific, cultural, economic, and political factors that drive disputes like his, environmental disagreements that pit citizens, businesses, and workers against the federal government and environmental nongovernmental organizations (ENGOs). Bundy's standoff in the Nevada desert is really just the tip of the conservation iceberg. Beneath the surface of the water lie grievances, disagreements, and divisions that fuel biodiversity conservation conflicts. Understanding these differences is the aim of this book. Unless we grasp the deeper reasons for why biodiversity divides us, such as how protecting a diminutive tortoise could go so wrong, we are doomed to keep setting the stage for fanatical confrontations over nature.

A common definition for biodiversity is the "full variety of organisms at all levels of biological organization, including the genetic, species, and ecosystem levels."[6] Many environmentalists believe that biodiversity is a proxy for planetary well-being. They see the growing extinction problem as justifying far-ranging, and sometimes radical, political changes. Biodiversity decline may eventually eclipse climate change as the flashpoint for environmental politics. It is poised to fuel partisan conflict for decades to come.

Why has biodiversity conservation been so divisive? How will it continue to divide us? What hope is there for a more productive kind of environmental politics?

DÉJÀ VU

Meeting in both Kunming, China, and Montreal, Canada, the world's nations recently adopted an ambitious new framework for conserving biodiversity. United Nations Secretary General António Guterres put the matter bluntly: "Let me be frank. Humanity is waging a war on nature." Other participants echoed this alarming language, citing biodiversity loss as "posing major risks to human survival" and emphasizing the need "for nothing less than urgent, transformative action."[7]

As we will see, the storyline of a coming conservation crisis doesn't resonate with everyone, probably not even a majority of people—including many scientists. What biodiversity is, why it matters, and what we should do to safeguard it are all spectacularly complex and highly disputed questions. Yet, to listen to national representatives at Kunming and Montreal, you wouldn't know it. The narrative dominating the UN agreement is one of existential stakes, self-evident policies, and a short window to avert catastrophe.[8]

The talks in Montreal came on the heels of a UN report estimating that some one million species are threatened with extinction. According to an analysis from the Stockholm Resilience Centre, humans have crossed multiple "planetary boundaries." We seem to be demanding more from the biosphere than it can sustainably provide.[9]

Corrective efforts look to be languishing. The world fell short of targets to protect 17 percent and 10 percent of the world's land and sea, respectively, set in 2010 in Aichi, Japan.[10] The Kunming-Montreal Declaration demands that we fulfill an even more audacious promise: protecting 30 percent of the world's land and seas by 2030.

Signatory nations must do far more than merely protect threatened ecosystems. The document calls for people to eventually be "living in harmony with nature."[11] According to some scientists, harmonizing with the environment doesn't just mean getting climate change under control. They contend that we must fundamentally change the way we produce our food and other resources, eliminate plastic waste, and control the spread of alien species.[12]

There are also demands for personal sacrifice, often singling out the people who live in the world's more affluent nations. Biodiversity advocates argue that people must learn to eat far less meat, have fewer children, and consume less stuff.[13] Calls to action have transformed into lectures about the need to reexamine not just our environmental beliefs and actions but also our very sense of self. As one representative from the Philippines put it, "Who are you, and are you who you need to be?"[14]

We are told that biodiversity progress is being stymied by a cadre of denialists, "industry shills," who are paid to sow doubt about the scientific reality of species extinction.[15] The editors of the prestigious journal *Nature* even invited ecologists to write a kind of "how to" guide for recognizing and countering different forms of "extinction denial."[16]

Are you feeling déjà vu? This all sounds a lot like the last three decades of conflict over climate change. And most of us would probably agree that climate politics has mostly been a gridlocked, unproductive mess. Of course, there's been some progress. Eighteen percent of the world's energy now comes from low-carbon sources. However, total fossil fuel consumption has yet to decline. And some international surveys have found that climate skepticism is on the rise, despite countless explainer articles and exhortations coming from some of the world's most famous scientists.[17]

There is good reason to worry that the political dysfunctions of climate change will also characterize biodiversity. Something about the last century's environmental thinking appears incompatible with the politics of twenty-first-century global challenges. But what should we be doing instead? How could we be *collectively wiser* this time around?

I believe that we can only begin to answer this question by better understanding why people disagree about biodiversity. Our lives and societies are everywhere defined by difference and discord. How should governments balance demands for equality with people's desires for individual freedom? What are the requirements for a good life? What makes for a good environment shouldn't be any different. If Americans can't seem to ever agree on the state of the economy, ongoing conflict in the Middle East, and countless other political issues, I doubt we'll ever achieve consensus on living in harmony with nature. Yet, even if unanimity were possible, it wouldn't be desirable. Democratic politics works best by transforming citizens' endless disputes into workable policy, not by trying to make disagreement disappear.

The story coming out of Kunming and Montreal, however, leaves little room for dissent. The crisis is self-evident, an indisputable conclusion drawn from following the science. The facts demand that we embrace new ecologically harmonious lifestyles. We must radically transform ourselves, and our societies, if human civilization is to survive. Skeptics and denialists will need to face the truth and be brought in line.

That story is built upon ideas that have long defined environmental politics, beliefs about the proper place of humans within nature and the rightful role of science within a democracy. If we are to learn how to channel our disagreements about nature into more productive political decisions, we will need to bring these assumptions out into the open and see them for what they are.

HOMO EXTINCTOR, THE EXTERMINATING SPECIES

Humanity "is a rapacious, ubiquitous, predatory . . . species." It is "a malignant . . . process engaged in the conversion of all plant, animal, organic and inorganic planetary material into human biomass."[18] In other words, people are a cancer on the planet. They grow incessantly and seem intent on putting the entirety of the environment in service of their own species' needs.

Readers have probably heard that argument before, although not necessarily from a hardcore environmentalist. Often it is said as a lament or thought-stopping cliché, uttered when the person you're having a beer with doesn't want to belabor the environmental challenges facing twenty-first-century humans, when they desperately want to talk about anything else.

Still, environmentalism has traditionally presumed that human society and the natural biosphere are profoundly incompatible. Eminent biologist E. O. Wilson went so far as to argue that "only by setting aside half the planet in reserve, or more, can we save the living part of the environment."[19] Saving the earth by reversing biodiversity decline, in this viewpoint, requires evicting people from much the landscape.

Writing in 1967, Arthur Koestler speculated that humans were "a biological freak, the result of some remarkable mistake in the evolutionary process."[20] Although his focus was people's tendency to go insane, many environmentalists would nevertheless see our current predicament as the product of a kind of madness, one that is uniquely characteristic of our species.

This madness features prominently in Kurt Vonnegut's novel *Galápagos*. As the narrator puts it, "The only real villain in my story [is] the oversize human brain."[21] People's "big brains" are the driving force in Vonnegut's book. And the three-pound organ generally gives bad advice. The novel's main characters, largely swindlers, greedy wretches, and morons, are led by their brains to commit ultimately self-destructive acts of vanity, deception, and cruelty. I won't recapitulate the book's plot, for it is largely beside the point. Suffice it to say, Vonnegut tells of the end of the human race. But he also describes its renewal through a band of survivors, who brave open waters to crash land at the northernmost island of the Galápagos archipelago.

People's "big brains" don't just drive human suffering in Vonnegut's novel, they also wreak havoc on "a very innocent planet."[22] Vonnegut presents the human species as insatiable, "growing for the sake of growth alone, and consuming all and poisoning all." People's insatiability degrades a "once beautiful and nourishing planet."[23]

Humanity's inherent perversity is represented by one of the story's main male characters, Captain von Kleist. He spends his final days stranded on Santa Rosalia, fantasizing about bringing the island's freshwater spring under his control. The captain, who doesn't actually know how to navigate a ship, dreams of dictating the spring's flow in order to prevent others from enjoying it without his permission. What stops him from inadvertently destroying the island's only source of drinking water is simply the lack of modern tools. As Vonnegut's narrator warns, "In this era of big brains, anything which can be done, will be done—so hunker down."[24]

The story's hero is nature, for it ultimately strikes back at humans' villainous brains. A disease emerges that renders most people infertile. A radiation-induced mutation causes the descendants of the small band of survivors to eventually evolve into seal-like creatures, whose "purely ornamental nubbin" fingers and shrunken brains are fortuitously unable to repeat the deviant ingenuities of the past. Humanity's fate is sealed when a great white shark swallows an advanced handheld computer called Mandarax, along with von Kleist and the story's female lead. One big bite severs the last remaining connection to the "big brain" era of human development.

Vonnegut presents people's brains as evolution gone awry. Although his book is a work of absurdism, it is anything but absurd to wonder about the potential maladaptiveness of human intelligence. Such questions are being asked by leading evolutionary anthropologists. Although evidence is limited, some researchers believe that human predecessors, and even Neanderthals, refused to migrate across open water and other large geographic barriers. It could be that "some freak mutation made the human insanity and exploration thing possible," speculates geneticist Svante Pääbo, which "changed the whole ecosystem of the planet and made us dominate everything."[25] Early *Homo sapiens* must have been a bit mad to have risked life and limb by venturing into the oceans, to attempt to discover islands that they couldn't see from the shore. It seems crazy to start a journey without being sure of your destination. Perhaps the gnawing urge to push boundaries in the never-ending search for the next frontier, an inextinguishable feeling of wanderlust and dissatisfaction, is something that only *Homo sapiens* are genetically predisposed to feel.

More optimistic observers see humanity's restlessness as a resource, believing that human ingenuity will lead us through the current dark age and into a "good Anthropocene." The future could be an era of "vastly improved technology and management, wise use of Earth's resources, control of human and domestic animal population, and overall careful manipulation and restoration of the natural environment."[26] Indeed, we'll examine proposals for delivering more vibrant ecosystems and a wilder planet later in this book.

Depending on whom one asks, the problem is either that humans intervene too much in the environment or that they have been intervening too little. Either we are destined for Faustian tragedy or we let our shyness about tinkering with nature get in the way of creating a world of sustainable abundance.

The former view certainly dominates environmental debates, especially climate change. Science is often called upon to authoritatively demonstrate the horror of humanity's unwavering penchant for environmental destructiveness. Scary-looking graphs and figures inform citizens of wealthy nations that they are exceeding planetary boundaries, that crossing ecological tipping points will make societal collapse all but inevitable.[27] Popular personalities on X, formerly known as Twitter, claim that "scientists say without an immediate shift away from capitalism global warming [is] so hellish that it's beyond the ability of humans to adapt."[28]

Digging deeper into the science, however, often uncovers pesky uncertainties and complications about the assumed incompatibility of humans and nature. Consider once again the tortoise at the center of the dispute between Cliven Bundy and the BLM. One would think that efforts in the 1990s to cut Clark County ranchers' grazing access were motivated by the unequivocal threat that cattle must have posed to the Mojave desert tortoise. But as late as 2002, a US Geological Survey report had concluded that "the evidence [of grazing impacts on tortoises] is not overwhelming."[29] Even if research results have since solidified against cows, how federal agencies decided to act against Cliven Bundy in the absence of conclusive knowledge is a clue that there is more to environmental disputes than just the facts.

As we'll see in later chapters, the idea that humanity's restlessness and ingenuity are inevitably destructive environmental forces doesn't exactly come from science. It is one possible story—one among many. Even though there is considerable evidence of the negative impacts of human societies, there is also reason to believe in nature's resilience and people's capacity for both adaptation and creative reconciliation. Many of the divides covered in the following chapters are founded on the belief that one, and only one, story about nature represents the Truth. Achieving biodiversity democracy will require being far more open to disagreement.

THE PROBLEM OF ENVIRONMENTAL DEMOCRACY

"I have a feeling that climate change may be an issue as severe as a war," notes public intellectual James Lovelock. "It may be necessary to put democracy on hold for a while."[30] Such sentiments are not rare. Some political philosophers argue that the urgency of global environmental challenges is incompatible with the slowness of democratic compromise.[31] Journalists speculate that meeting the demands of climate change requires "new ways of closing the gap between politics and science, by any means necessary."[32]

Environmentalists' traditionally dismal view of humanity sets them up to see science as nature's savior. If the world is despoiled because of people's failure to clean up after themselves, then designating a knowledgeable babysitter for the planet's eight billion unruly and ignorant children seems like a self-evident solution. Restoring ecological harmony, thus understood, requires establishing a clear and authoritative sociopolitical order, the rule of *natural* law.

So far, no government or organization has wrestled control from democratically elected politicians to "save the planet." However, scientists involved with the Intergovernmental Panel on Climate Change (IPCC) have recently proposed that *they* should prescribe the policies to correct the climate situation and perhaps even have the authority to oversee their implementation.[33]

And a well-funded movement in the United States wants to get the American national energy system declared unconstitutional, which would force an immediate ban on all fossil fuels, regardless of the human consequences and despite whatever citizens, Congress, or the president think about it.[34]

At the same time, "Follow the Science!" has become a pervasive political slogan. It presumes that issues such as climate change and biodiversity decline are primarily solved by listening to experts. That assumption also makes the opposite true: Failing to implement the right policies must be the product of willful ignorance. Indeed, protestors in North America and Europe now regularly glue themselves to roadways or to priceless works of art under the impression that, if governments would only heed the voice of science, environmental progress could finally be achieved. Politics, in this view, isn't how we settle public disagreement. Rather, it is a barrier to doing what some environmentalists consider to be self-evidently correct. Compromises, especially with industry, are always suspect, derided as selling out the truth to appease backward citizens or illegitimate political interest groups.

This is exactly how journalist Christopher Ketcham depicts the conservation conflict over the sage grouse, a striking chicken-sized bird that is native to sagebrush landscapes throughout the United States and Canada. In Ketcham's view, stringent restrictions on cattle grazing and oil and gas drilling were rejected by politicians primarily because their careers depended on "fellating" these industries. The result is the continued "spoliation" of nature. The right thing to do is ostensibly crystal clear to all "researchers not in the pay of [these] companies," but unfortunately doing so is being derailed by "politics."[35]

As I described in my last book, *The Divide*, this attitude is characteristic of our hyperpolarized political era. Partisans of every stripe see themselves as on the right side of history. They think that the causes of the world's problems (and the best solutions to them) are self-evident. They believe that disagreement can only be explained by ignorance, corruption, propaganda, or a pathological refusal to respect science. *We* know the facts; *they* are adrift in post-truthiness and fake news. This line of thinking doesn't quite go so far as Lovelock's, but it does share his fundamental mistrust of democracy and popular dissent. It reflects the belief, as political scientist Jeffrey Friedman put it, "that those who claim to see things differently *must* be lying."[36] The implication is that if corporations didn't meddle in politics and if citizens didn't consume misinformation, then democracies would have long ago arrived at a rational consensus for solving contentious environmental issues.

Scholars have, perhaps inadvertently, fed into this narrative. *Merchants of Doubt*, a book by historians Naomi Oreskes and Erik Conway, describes how contrarian skeptics exaggerated the relevant scientific uncertainties regarding cigarettes and climate change in order to delay restrictive regulations.[37] There is no doubt that lobby groups, whether motivated by financial interests or by the belief that regulations represent "creeping socialism," regularly question the validity of university and government research. Professional climate skeptics make unorthodox claims, such as connecting climatic changes to shifts in solar activity or Earth's orbit, and sometimes have even spread false accusations about prominent scientists.

However, readers of Oreskes and Conway's book too easily come away with naive ideas about how politics works.[38] *Merchants of Doubt* gives the impression that the right policies don't get implemented mainly because of doubt merchants' corrupting influence. If only American journalists refused to tolerate arguments from industry or prominent skeptics who stray too far from the scientific consensus!

Even if everyone agreed that people are driving climate change, thorny uncertainties would remain. Exactly how much of it can be attributed to humans? How warm will the planet get? What precisely will the impacts be on coastal cities or arid regions? What is the right balance of prevention and adaptation to climatic changes? How should we prioritize different potential solutions? Solar? Nuclear? Carbon capture? Geoengineering? What is the best combination of restrictive regulations and policies to foster more innovation? By focusing narrowly on highly visible efforts to sow doubt about basic climate science, Oreskes and Conway miss the forest for the trees. They understate the complexity of the situation and thereby exaggerate the potential for political consensus.

The speed of policy progress usually isn't limited by scientific agreement but rather by practical challenges. Officials face difficult questions. How can competing economic interests be balanced? How can regulatory, infrastructural, and other barriers to change be overcome? How can the right kind of innovation be effectively spurred? And how can policies be developed that the public can get behind? Change can be politically infeasible for all sorts of reasons. Nevertheless, the underlying narrative of *Merchants of Doubt* reduces the problem to citizens being conned into denying the mainstream factual consensus.[39]

Unsurprisingly, "sowing doubt" has become a trope among environmentalists. And it is used to prop up a simplified, *Star Wars*–like story of environmental politics, one where there is only a science-denying corporate empire and the fact-respecting eco-rebels. A recent article about the challenges of passing stricter European Union (EU) regulations on pesticides, for instance, presents industry's complaints as inherently illegitimate. Agribusiness disagrees with strict, legally mandated cuts to chemical use, expressing concerns about potentially increasing Europe's reliance on agricultural imports. We are told that these statements are nothing but "delay tactics," that industry-funded analyses are intrinsically suspect. And the article presents environmentalists' views as rooted in unequivocally authoritative science. Worries about food security, for instance, are mere "scare stories" because we have more than enough food. But the cited studies include an unmentioned caveat: There's plenty to eat only if "there is radical societal adaptation" to plant-based diets.[40]

There are, no doubt, good reasons for reducing pesticide use, and arguably the agro-chemical industry has too much influence in governmental negotiations. Democratic principles, however, require that nearly everyone gets to influence rulemaking. This includes the industries that bear the costs of compliance, even if they invariably fight to water down the recommendations of university scientists and environmental campaigners. It isn't exactly sowing doubt for agribusiness to disagree about how to balance the potential impacts on food security and national trade with diet and climate change. Any opinion on the matter depends on both relevant, but disputed, facts and subjective values. Ultimately, the "follow the science" mantra makes it hard to accept the inconveniences of democracy, for it drives outrage and disappointment over the reality that politics is, well, full of politics.

In order to avoid repeating the pathologies of the climate change debate for biodiversity, we will need to dispense with simplistic narratives and honestly try to understand the underlying causes of contention. For instance, journalist James Pogue's book on Cliven Bundy uncovers a far more complicated story than a rancher flouting BLM rules and conservation science. To begin, grazing restrictions often put ranchers out of business. And these specific rule changes came as part of a deal to remove boundaries to Las Vegas's continued growth. Ranchers, quite reasonably, saw themselves as being sacrificed for the sake of suburban expansion, which arguably posed a much bigger threat to desert tortoises than cattle.[41] The case of Cliven Bundy ends up being surprisingly complex. If it can't be honestly depicted as a straightforward battle between science believers and denialists, then what about the rest of biodiversity conservation?

That does not mean that Cliven Bundy should be allowed to keep running his cattle, nor that his viewpoint is correct. He often talks as if he had a constitutional, and perhaps even God-given, right to the landscape around Gold Butte National Monument. Bundy seems to believe that Americans east of the Mississippi should have little say over what he does on public lands in Nevada, even though federally funded waterworks are what makes Western agriculture possible.[42] Bundy is as fanatically convinced that he is right as many of his enemies within the environmental movement.

Democracy falters when politics becomes defined by dogmatic righteousness. Political disputes become nasty, and differing perspectives turn into entrenched ideologies. A necessary condition for finding our way back to a strongly democratic form of environmental politics is to restore our sense of curiosity. Why is it that reasonable people can see the same science, look over the same landscapes, and experience the same nature but believe very different things about conservation?

Understanding alone is probably insufficient. A vital piece of the puzzle is figuring out how we can more reliably come to productive compromises. That's not as easy as it sounds. If we accept that there are plenty of good reasons to disagree about biodiversity conservation, then we shouldn't expect to ever achieve political consensus regarding how to live in harmony with nature. If we are to realize more biodiverse landscapes, political progress will need to happen despite the persistence of heated conflicts over what a good environment looks like. Fortunately, partly reconciling the political demands from disparate factions is exactly what democracy can accomplish, if we let it.

A ROAD MAP

Because I wanted to write a book that nonacademics would want to read, what follows is not a systematic philosophical treatise.[43] Instead, I have tried to show more than tell. I have described events, experiments, and disputes involving biodiversity in order to illustrate the major conflicts regarding conservation. Two main questions will guide this journey: How should people fit into nature? And what is the place of science in a democracy?

I won't survey too many examples of corruption and political intrigue, or times when the cries of business interests unfairly drowned out all other perspectives. Such cases no doubt exist. However, one of my goals is to try to charitably understand the people who find themselves on the opposite side of traditional environmentalists. I want to depict how differing views of biodiversity conservation are reasonable. I do this in order to break out of simplistic depictions of opponents as ignorant, as manipulated by corporate interests, or as simply out of touch with changing times. We will only learn to have more productive democratic disagreements about biodiversity by treating our fellow citizens with empathy, even if we have grave doubts about their beliefs and intentions.

The next chapter begins with what journalist Keith Kloor termed "the battle for the soul of conservation science,"[44] a dispute launched when some biodiversity scientists called for a radical departure from traditional conservation. The following cases all relate to some facet of the claimed biodiversity crisis. The topics are science heavy and complex. I have tried to keep jargon to minimum while providing enough detail to show why authoritative scientific answers are rarely forthcoming. We will see why the dominant biodiversity story, one predicting impending ecological catastrophe, is unavoidably as much politics as science.

Chapter 3 digs deeper into a particular disagreement within biodiversity: what to do about alien species. Invasion biology treats them as ecological contraband, as harmful until proven otherwise. Not only is the concept of invasive alien species not as coherent of a scientific idea as one might think, scientists also disagree about how to respond to the risks posed by newly arriving species.

Yet, biodiversity conservation isn't all extinction and degradation. Chapter 4 surveys a relatively new area of research and practice called rewilding. It offers both a more optimistic narrative about people's role in biodiversity and a hopeful vision for bringing about a wilder world. Like invasion biology, however, the story of rewilding seems to pine for places without people—a feature that limits its ability to speak to those who live from the land.

The following chapter focuses on the practical sources of conservation conflicts. Differences in scientific opinion are important, but they aren't really what makes biodiversity contentious in places such as rural Idaho. Environmental disputes have long histories. They are rooted in differing ideals regarding how to live on the land and who gets to decide what the landscape needs. The battles are frequently acrimonious because they are rarely resolved democratically. Environmental politics too often pits rational visions for ecological harmony against the people left out of those visions.

But all is not unwell. Chapter 6 surveys efforts to reconcile these conservation conflicts. How can economic development be more often squared with protecting endangered species and safeguarding habitat? How can protecting nature or introducing species be less disruptive for rural populations? How might we feed the world while also giving more room for nonhuman nature? None of the examined solutions are panaceas. However, taken together, they illustrate a promising alternative to "follow the science" for trying to productively work through polarizing environmental disputes.

The penultimate chapter looks at more radical conservation proposals. Ideas such as de-extinction, assisted migration, and facilitating the evolution of threatened species may be necessary, in light of the pace of climate change and other environmental challenges. Too often the debate is preoccupied with trying to decide in advance whether we *should* use these technological approaches. The more useful question is how we might more quickly learn if and how we *could responsibly* test and then implement them. Humanity is already and always engaged in a collective experiment in ecological living.

Slowing and reversing biodiversity losses will require learning by doing, and a range of strategies can ensure that we do so more swiftly and prudently.

In the conclusion, I offer my vision for how people are to dwell responsibly on the planet. I sketch the beginnings of a positive role for humans in nature. I chart a more helpful place for scientific experts in environmental disputes. I make a plea for a kind of environmentalism that includes nearly everybody. And I describe how democracy can help us realize more intelligent and sustainable societies.

WHAT IS BIODIVERSITY?

The following chapters make use of several concepts from biodiversity science. Knowing their meaning isn't essential for understanding my main arguments, but it helps for following the scientific debates. *Ecosystem* refers to the web of relationships between organisms and the physical environment or the *habitat* they find themselves in. I will use the word *landscape* to refer to physical territory made up of many kinds of ecosystems, including deserts, grasslands, and forests. Species occupy different ecological *niches* within a habitat, where they thrive and maintain a viable population. And they typically are engaged in some set of *ecological functions*, such as pollination, spreading seeds, recycling wastes, and filtering water.

The most important scientific concept in this book is biodiversity. As mentioned above, it encompasses the variety of life on Earth, from genes to ecosystems. That sounds great, but scientific concepts also have to be measurable. Measuring biodiversity is not so straightforward. Take a plot of land. You could measure tree biodiversity by simply counting the number of species on it (i.e., species richness). But wait, what if the land is a patchwork, where each corner is occupied by a different set of species? That doesn't seem very diverse. Therefore, we should also measure the evenness of species. How well interspersed are they? But wait, what if all the trees on a plot are evergreen species? Isn't that less biodiverse than one that also features deciduous (i.e., leafy) trees? So, we'd also want to take into account genetic diversity too. And, even still, it seems important that there are species that support a diversity of ecosystem functions.[45]

The situation becomes even more complex at different geographic scales. Scientists measure biodiversity for single habitats, between habitats, and across entire landscapes. Boreal forests, for instance, are fairly uniform, containing around twenty tree species. In mountainous tropical regions, you encounter vastly different communities of organisms simply by going up or down several hundred feet in elevation.[46]

No doubt this is all very scientific stuff. Measuring and counting has been fundamental to the biological sciences at least since Alexander von Humboldt traveled to the Americas and obsessively documented everything he saw. Nevertheless, we're ultimately faced with the question of what these measures mean for us. Why should we care about biodiversity?

Humans derive a lot of value from nature. It provides what are called *ecosystem services*. Rare and beautiful species leave us in awe. Insects and other species pollinate flowers and other plants we enjoy. Wetlands filter surface waters. Grasses feed our livestock. And our mental state often improves just by being in natural or wild spaces.

Research shows that biodiversity brings benefits. For instance, field experiments find that plots of land with more biodiverse vegetation are more drought tolerant and productive (i.e., there is more vegetation growing above ground level).[47] Yet, it is debatable that biodiversity per se defines nature's value, that it forms the basis for human survival, or that maximizing global biodiversity should be the goal of conservation.

Environmental philosopher Donald Maier spends hundreds of pages of his book, *What's So Good About Biodiversity?*, surveying the research and concludes that "the evidence and reasoning on offer . . . provide no credible grounds to suppose that nature-as-biodiversity is particularly valuable."[48] Apart from finding a number of logical and empirical inconsistencies, Maier mentions some more commonsensical objections. Many of the landscapes that we value and benefit from, such as glaciers, wetlands, and boreal forests, aren't especially biodiverse. Most of the gains that humans get from nature come from the relatively few species that we have domesticated, such as honeybees. In some cases, especially for agricultural pests or the viruses and bacteria that make people sick, we would prefer less diversity, not more. And finally, even if we can find benefits from local biodiversity, it isn't evident why diversity on a *global scale* should matter. We haven't been able to even locate most species, probably because many of them are tiny populations clinging to life in isolated patches of tropical forest. How could we possibly know that these species are essential for planetary health?

Having spent the last few years obsessively reading and thinking about biodiversity, I can't help but agree with Maier, at least partly. Compared with climate change, the cause-and-effect relationship between biodiversity and human well-being is more complex and uncertain. There is a relatively clear threshold of heat and humidity where a human body struggles to cool itself. We can imagine droughts and other extreme weather events becoming so severe that they make agricultural exceedingly difficult. However, we don't exactly know what the loss of even one million species would bring.

The rest of this book will still refer to biodiversity, under the assumption that there is some imperfect relationship between the diversity of species and nature's value. Readers, nevertheless, should keep in mind that it might be "wildness" or something else that we really care about in some situations. My wife, for instance, delights in leaving out food for songbirds and hummingbirds, and also for feral cats. It isn't really supporting biodiversity that motivates her, especially given that house cats don't help songbird populations. Rather, it stems from a desire to engage with the nonhuman world. Perhaps the point of environmentalism should simply be to multiply our encounters with nature and make those engagements more mutually beneficial.

People generally like environments that feel alive. We promote ecological aliveness by trying to better fit ourselves into a broader natural world not of our own making, rather than attempting to always force nature to submit to our desires.[49] What matters most may not be what exactly happens with the global catalog of eight million species. Rather, a better task for us humans may be to keep making a bigger effort to adjust our activities so that they are less disruptive to the life processes of nonhuman species.

Fortunately, we really don't need total philosophical clarity to go on this journey. I have actually come to believe that it is better if we don't try to rigorously define what a better Anthropocene looks like from the outset. What makes for a wilder planet Earth (and where humans fit in) is something we should be open to discover and to continually revise as we experiment with new ways of living.

TOWARD BIODIVERSITY DEMOCRACY

I wrote this book because I am a nature lover. Hiking, mountain biking, rock climbing, and camping are some of my favorite activities. I regularly hop off my bike on the way to work in order to try to identify different birds or a cool-looking insect that I have never seen before.

I am also a partisan for a wilder planet. My daydreams sometimes involve imagining buffalo or even camels and elephants eventually dotting my local New Mexican landscape. Such a vision may be an overly romantic and totally infeasible fantasy, but the best daydreams usually are.

As I began researching for this book, I started to worry that my preconceptions, as an environmental leftist, prevented me from understanding why conservation progress was so paltry. Years earlier, I had already become fascinated with how people arrived at starkly different views regarding science and technology. Why did some citizens mistrust nuclear power while others saw it as an environmental savior? How come climate change debates devolve into dueling studies and incompatible interpretations of the data? In the process of researching these cases, I came to the conclusion that people usually misuse science in heated political conflicts. The facts are often a distraction, or even a political weapon, used to obscure the concerns and values that really founded citizens' disagreements.

Then the pandemic happened. Like other Americans of my educational, economic, and political class, I dutifully embraced most viral precautions. I wore a mask. I got vaccinated multiple times. I allowed students to Zoom into class if they had a sniffle. Yet, I understood where COVID skeptics were coming from, that they had reasons why they mistrusted the Food and Drug Administration or Anthony Fauci.

Watching my fellow citizens openly mock the vaccine hesitant as they lay dying in hospital beds or respond with a shrug when blue-collar workers lost their jobs because of vaccine mandates, I was horrified. That US policy only counted vaccines as evidence of immunity, and not prior infections, left me perplexed because they were considered equivalent in Europe. I watched as state governors worked hard to reopen restaurants, bars, and major industries but prohibited gathering in churches, kept playgrounds chained up, and barred people from visiting loved ones in the hospital. The defense of these debatable decisions as scientifically ordained, as just "following the science" and therefore not also politics, was maddening. It revealed a lack of respect for ordinary citizens. And it ran counter to what I thought democracy meant.

Although I didn't personally agree with anti-vaxxers during the pandemic, I recognized that achieving good public health outcomes was impossible without trying to understand them. Otherwise, there was little hope for persuading them to change their mind, or at least to hammer out imperfect compromises. Because of governments' ongoing failure to understand them, Americans now face declining vaccination rates among five-year olds and lessening public support for mandatory shots for schoolchildren more generally.[50] More than a quarter of Americans now say they have little trust that scientists will act in the public interest.[51] "Follow the science" politics didn't lead to consensus. It has arguably left the United States even less capable of battling future pandemics.

This outcome is only part of the reason why claims about democracies now being imperiled by a post-truth age are profoundly misguided. They're also historically naive. America's first president, George Washington, left office decrying fake news. He lamented the gazettes' "ignorance of facts, and malicious falsehoods," blaming them for destroying the public's confidence in his leadership.[52] We should also remember that citizens could scarcely find a newspaper that wasn't full of outright lies until the mid-twentieth century.

Politics has always been defined by the spreading of innuendo, mistruths, and selective readings of facts. And people of all political stripes have acted in "post-truthy" ways. A nuclear research reactor was closed at Brookhaven National Labs in the late 1990s, for instance, after a relatively harmless leak of radioactive water.[53] Behind the effort to close it was an activist group led by Hollywood actor Alec Baldwin, who repeated questionable claims about both the leak and the laboratory. Another member, Helen Caldicott, even openly implied that lab officials were intentionally releasing radioactive waste and protecting themselves with special pills. The group later wheeled out a cancer-stricken boy on *The Montel Williams Show*, dubiously claiming that his illness was caused by Brookhaven. Contrary to the belief that post-truth only emerged with the Trump administration, this effort was spearheaded by Democratic Party donors.

We should not, however, conclude that something called post-truth is to blame for our political dysfunctions. Brookhaven National Lab's reactor wasn't shuttered because of misinformation but rather because the lab had betrayed the public's trust and failed to regain it. Brookhaven workers had irresponsibly managed both chemical waste and nuclear material for decades and then had the gall to dismiss local residents' concerns about toxicity for being insufficiently fact based.[54] When people aren't sure who to trust, expert opinion looks a lot less certain (and conspiratorial claims can seem a lot more believable).[55] In the case of Brookhaven, the science was relatively straightforward. Things get far messier when even scientists themselves disagree about the facts.

Regardless of what science is involved, the preoccupation with misinformation and post-truth blinds us to what lies beneath seemingly factual disputes, to the factors really driving heated political conflicts. Trust and mistrust, worldviews and values, and the suspected consequences for the people, places, and animals we love are usually far more decisive. They have more influence on how we view contentious public issues than whether we have read, understood, and believed the latest scientific study.

Both my research into scientific controversies and my frustration with how the pandemic turned out have shaped how I have presented conservation conflicts in this book. I used to believe that I could figure out the right answer for contentious issues. Now, I am more interested in how we could achieve durable outcomes that work for most everybody, despite their disagreements. This is exactly my hope for biodiversity—that we can create a wilder world while also avoiding turning ever more people into conservation enemies.

Cliven Bundy doesn't feature in the rest of this book. I began with his story because it illustrates one potential future for biodiversity, one that we ought to try to avoid. The fact that scores of men were willing to lay down their lives in a shoot-out with federal agents, all over a few hundred head of cattle, signals that something has gone very wrong with America's environmental politics. The event isn't something we should dismiss as just some fanatical right-wingers cosplaying as militiamen in the desert. Bundy's standoff presages the high-stakes confrontations awaiting us as scientists, activists, and international organizations call for transformative changes to save the planet. It might be possible to live more harmoniously with the nonhuman natural world without actively deescalating and depolarizing biodiversity politics, but I doubt it.

2 POLITICAL SCIENCE

It was a tense meeting, an unexpectedly fierce conversation for a group of scientists. Much of the contention was spurred by a single question: What was the future of biodiversity?

They had convened at the Virginia headquarters of TNC, one of the largest environmental nongovernmental organizations in the world. Michael Soulé, considered by some to be the "grandfather" of modern biodiversity conservation, was leveling serious charges against TNC's head scientist, Peter Kareiva. Soulé accused him of "demeaning national parks" and "doing a lot of damage" to the main scientific and political movement to protect biodiversity. Kareiva had recently published a contentious article that had offered a new vision for environmentalism. Soulé felt that it put conservationists under attack—that it amounted to saying, "Biodiversity is shit."[1]

"I'm not a biodiversity guy," responded Kareiva, "I'm an ecologist." He believed that the exclusive focus on species numbers led conservationists astray. Kareiva's "new conservation" included partnering with corporations. He advocated finding environmental solutions that worked for a wider spectrum of people, not just the minority of eco-warriors who sought to protect nature in fenced-off preserves. As a *Nature Conservancy* magazine profile put it, Kareiva wanted "conservation of the people, by the people, for the people."[2]

To Soulé and others in attendance, business involvement sounded more like corporate dominance, and TNC's "bashing" of traditional conservation had spoiled the efforts of other environmentalists. The meeting ended somewhat amicably, but Reed Noss, of the University of Central Florida, promised in a follow-up email to continue fighting the organization's "schizophrenic anti-biodiversity, anti-protected-areas rhetoric."[3]

The group did fight, albeit in the way that academics do: They wrote a book. *Keeping the Wild*'s recurring theme was denial. "Neo-greens" like Kareiva failed to see the magnitude of the problem and the scale of environmental degradation and believed too much in the power of technology. They favored human dominion over the environment, denying the need to alter human populations and lifestyles to match planetary boundaries. "Old" conservationists also wrote letters to the editor in major journals. One instance started off with a questionably tasteful metaphor, comparing the new conservationists with prostitutes and chiding them for mistakenly believing that seeking attention was the same as being desirable.[4]

In the dispute between Kareiva and conservation's old guard, science starts to look a lot more like politics. People fracture into factions, rhetoric becomes heated and accusatory, and questions regarding the proper values, goals, and means take center stage. As we'll see throughout this book, this is normal for environmental issues. Conservation conflicts, however, have become especially contentious as some warn of an impending extinction crisis. There is a renewed sense of urgency. Many people believe that we're living through the Anthropocene—a special geological epoch characterized not merely by widespread human impacts but also by the risk of self-annihilation. It was into this turbulent situation that Canadian ecologist Mark Vellend submitted a paper to *Nature*, one of the world's preeminent science journals.

As journalist Keith Kloor reports, Vellend's argument made the first reviewer very uneasy.[5] The study was a meta-analysis—research that collects and synthesizes dozens or even hundreds of findings from other scientists—and it presented results that contradicted a common conservation assumption: that global declines in biodiversity are matched by similar reductions at a local level. If planet Earth has fewer species, then the same should be true for the state of Minnesota and even for a patch of forest on the coast of Lake Superior. However, Vellend's paper found the opposite relationship for plants, and a similar analysis led by Maria

Dornelas seemed to confirm this result for animals.[6] As species died out, newcomers filled the vacuum. And while some field sites included in the analysis experienced dramatic declines, others became more biodiverse during the period of study. The reviewer demurred, warning that the media could latch onto Vellend's paper and lead people astray. They argued, "It is important not to pave the way for that conclusion [that biodiversity isn't declining] to be reached by the public."

Nature rejected the paper. Although the team ended up publishing the study in *Proceedings of the National Academy of Sciences*, the reviewer's words stuck with Vellend. He found similar sentiments among audience members at conferences and online. The austere language of scientific critique quickly gave way to personal attack, including the claim that Vellend had an "agenda" that clouded his judgment and that he produced "foul science."[7]

It is understandable to want science to be our North Star, guiding us as we grapple with an apparent crisis in species conservation. However, biodiversity is not merely a scientific matter. Since the very beginning, it has been shaped by a contentious question: What story should be told about nature and who gets to tell it? The stuff of politics (e.g., different visions of the good life and society, ideas about risk, rights, and responsibility) are unavoidable when talking about biodiversity. Science cannot rise above it all. Moving toward a more productive and less polarizing kind of environmental politics will mean honestly reckoning with these complexities rather than wishing them away.

IT'S COMPLICATED

As I have combed through books and journals, the complexity and multiplicity of environmental science has often overwhelmed and confused me, especially where it intersects with policy questions. And I am not the only one. A hydrologist friend of mine once told me about a time he served as an expert adviser for a public hearing. The focus of the meeting was a proposed natural gas drilling project. After sitting through hours of conflicting presentations from scientists allied with various interest groups and governmental bodies, one exasperated public official got up and declared, "Someone here . . . is full of baloney!" He hoped that the science would have been able to provide clear and unambiguous advice, a single authoritative answer, even for contentious questions. If only.

The debate surrounding Vellend's claims of "no net loss" in local biodiversity illustrates how seemingly straightforward scientific questions only seem to become more complex the more scientists try to answer them. Criticism of the article by McGill University scientist Andrew Gonzalez focused on a number of limitations.[8] The sites surveyed for the study were unevenly spread across the globe. North America and Europe (places with the most research money) were overrepresented. Critics also questioned whether the surveys were long enough. Sometimes, extinctions take time. Scientists liken the process to a growing debt, where the effect of impacts and habit loss snowball and result in species disappearance only after some delay. Population numbers can deceive us because some organisms may be "dead species walking."

I won't try to relay all the details of the debate.[9] Suffice it to say, the back and forth revealed methodological and practical limits found in a lot of biodiversity research. Nearly all global studies suffer from the same geographic biases (i.e., too little data from outside North America and Europe). Although Vellend and his coauthors didn't believe that combining short-term studies prejudiced the outcome of their research, other scientists sided with the critics. Skeptics argued that many of the included studies were too different from one another to be part of the same meta-analysis. It remains disputed exactly when combining local, short-term studies can actually provide a convincing argument regarding global biodiversity changes. A recent paper concluded that reliably detecting such changes would require a network of thousands of observation sites randomly distributed across the globe, which measured local species richness with little to no error.[10] But such a network does not currently exist.

In any case, Vellend's and Dornelas's research doesn't stand alone. Others have found a similar pattern. For instance, reptile and amphibian diversity is higher today in California. A larger variety of fish swim in the rivers of many islands. A greater variety birds or mammal species can be found in many regions. That said, nearly all scientists acknowledge that the *composition* of species has changed.[11] For example, some 140 species of reptiles have been introduced to South Florida, including Burmese Pythons.[12] Some of these introductions displaced already-existing species. But what should we make of such changes?

Imagine a forest. As globe-trotting generalist species replace unique local plants and animals, it starts to look no different from similarly positioned landscapes across the world. Local species diversity remains flat or increasing while global biodiversity declines. The eventual result, in other words, could be an Earth-wide simplification of nature.[13]

Some think that this potential outcome would deplete nature's resilience. Just as a worker will have a variety of stocks and bonds in their retirement fund in order to weather whatever market disruptions occur, the "insurance hypothesis" posits that high levels of locally distinct biodiversity stabilize ecosystems.[14] Without unique species, there is a loss of genetic resources for living things to adapt to disruptions. On the other hand, the species replacing locally extinct varieties have been able to do so because they are already highly adaptive generalists. How do we know that these species won't actually better weather future climatic changes?

A bigger concern about Vellend's study had to do with what his team purposefully left out. They focused on how biodiversity has changed over time for intact areas. So, if a research site was turned into an Arby's parking lot, for example, it didn't factor into their results. Such land-use changes are studied by other scientists, such as University College London ecologist Tim Newbold, who compare the biodiversity between "natural" habitat and "converted" land. Unsurprisingly, studies in this area find that fewer species live in cornfields, pasture, and cities, estimating that the decline in species richness could be up to 77 percent in the worst cases.[15] Their concern is that Vellend's and Dornelas's studies miss the bigger picture: Around a million hectares of forest is cut down every year in Brazil alone, and urban areas are expanding around the world.

It is tempting to try to pick and choose who is really right in this debate, but that task is beyond my expertise. My aim is merely to show how science really works. Media representations, and often our own imaginations, are dominated by heroic images of white-coated scientists. One envisions new research results unequivocally exposing ignorance and providing authoritative facts on the nature of reality. But studies don't really *answer* so much as produce additional questions. Partly solving one problem just brings more uncertainties to the surface, exposing new mysteries and often raising doubts about previously settled research.

In demonstrating this very point, science policy scholar and geologist Daniel Sarewitz is fond of citing Yucca Mountain. The area was supposed to house America's high-level nuclear waste. But its suitability hinged on estimates of the "percolation flux" of water through the local geology. Basically, is it dry enough to trust that nuclear storage canisters won't eventually rust and then leak waste into the environment? These estimates, however, varied by almost three orders of magnitude, from a tenth of a millimeter to four centimeters per year, a difference that matters a great deal when waste needs to be stored for thousands of years. Scientists tried to reduce uncertainty by testing the bedrock for a rare isotope of chlorine that is produced by nuclear testing. The question was seemingly simple: Can we find evidence of sixty-year-old nuclear weapon tests at a depth of three hundred meters? If high levels of this isotope were detected, then they would know the rate that water flowed through the rock. But coming up with sufficiently precise and accurate measures of this chlorine isotope ended up being very challenging, with scientists from different laboratories producing divergent results. Attempting to answer a question about percolation flux didn't reduce uncertainties. Rather, it raised still more complex questions about how best to estimate water flow by measuring chlorine.[16]

The widespread belief in the myth that science produces unequivocally authoritative knowledge leads us astray, to forget that Truth often remains just ever so slightly out of reach of researchers. Criticism of Vellend's and Dornelas's work doesn't make their claims of stable levels of local biodiversity wrong any more than their defenses vindicate them. Rather, the question was whether the data really allowed them to make such strong conclusions. And the debate raised further concerns about producing better studies, refining data analysis techniques, and bringing extra care to meta-analyses. For what it's worth, an editorial by Boston University biologist Richard Primack concluded by praising the work by Vellend and others for pushing the field forward and questioning conservation dogma, even given their studies' limitations.[17]

Vellend's research has had a significant impact on the field. When I have spoken to biodiversity scientists, many of them credit Vellend and others for the fact that the phrase "biodiversity change" has supplanted "biodiversity loss." Similar to the transition from "global warming" to "climate change," the narrative surrounding biodiversity is becoming more complex. No longer can it be reduced to a story of decline. But that transition isn't going to be a smooth one, for biodiversity conservation isn't simply academic research. It is deeply intertwined with ideas about nature itself.

THE QUANTIFICATION QUANDARY

Yale political scientist Charles Lindblom was still in the early years of his career in 1954 when he was invited to study at the RAND Institute. RAND is a think tank, which was founded at the end of World War II to help guide federal research and development decisions. Lindblom was tasked with the "criteria problem." Say you wanted to know how to pick the most effective method to bomb the Soviet Union with nuclear warheads (an actual question studied at RAND). Would it be better to maximize the damage inflicted per dollar spent or minimize the aircrew lives lost per amount of damage inflicted?[18] Which criteria helped ensure the best outcome when confronted with that complex choice?

The criteria problem proved to be formative for Lindblom. Although known today for his contributions to political science, he originally trained in economics at the University of Chicago, a field that had become increasingly captivated by the power of quantification. Similarly enthralled with numbers were RAND Institute researchers, who would later have a big influence on the Kennedy administration. Then–Secretary of Defense Robert McNamara was a fan of systems analysis and operations research, having used mathematics to optimize automobile manufacturing supply chains and other processes at Ford. With the help of RAND, McNamara marshaled a whole array of statistics (e.g., body counts, communist attacks) to convince himself that the United States could win and was already winning the war in Vietnam, at least according to the chosen criteria.[19]

An old apocryphal joke tells of generals trying to predict the end of the conflict, feeding a Pentagon computer with reams of data and heading home for the weekend. Upon returning to their offices on Monday morning, they were greeted by a single punch card alerting them that they had already won the war in 1965.[20]

Lindblom had recognized the fundamental hazard in the whole approach early on. In the words of historian David Jardini:

> Despite the mathematical sophistication . . . the criteria upon which RAND systems analysts based their conclusions were heavily value-laden and were typically selected according to the personal preferences of the analysts. The search for optimal solutions to complex policy questions was thus a trap since *the more complex the systems analysis became, the more deeply biased the analysis became.*[21]

How many Vietnamese and American lives would have been spared had McNamara recognized the biases in his models? Although the rest of us aren't conducting a war in Southeast Asia, we still ought to be wary of quantification. Modeling is probably both necessary and unavoidable. Nonetheless, models sometimes say far more about the modelers than about the nature of reality because they ultimately define the model's criteria based on *what matters* to them. As a result, the modelers can end up laundering their preferred values "behind the façade of scientific neutrality." They deceive not only themselves but everyone, leading us to falsely believe that an objective analysis of the facts compels a particular response to public problems rather than some people's political worldviews, desires, and anxieties.[22]

THE MEASURE IS THE MESSAGE

"Humanity has wiped out 60% of animal populations since 1970, report finds." So reads the title of a *Guardian* article on biodiversity loss.[23] There are dozens like it, and most of them cite the Living Planet Index (LPI). The biodiversity indicator, which is managed by the Zoological Society of London and the World Wide Fund for Nature, is not really a straightforward measure of wildlife's well-being, even if many journalists take it to be one.

The LPI has declined more than 60 percent since 1970, but that doesn't literally mean that sixty out of every hundred of the world's wild animals have disappeared.[24] The LPI is an average calculated from a small sample of the world's species, some five and a half thousand out of an estimated total of five to ten million species. It also averages those losses in an unusual way.

Most of us are familiar with the arithmetic mean, where you add together a set of numbers and then divide by the number of figures that you have. Let's say we had four roughly equal populations of animals, starting with a hundred of each, of which three were relatively stable but one crashed to 1 percent of its original level. The arithmetic mean of the new populations would be 110 + 90 + 100 + 1 divided by 4, or roughly 75, implying an average decline of 25 percent.

The LPI doesn't use the arithmetic mean. Rather, it uses a geometric one. This requires multiplying the four numbers together and then taking the fourth root of the result. Don't worry if you don't remember these calculations from high school. The point is that the geometric mean equals about thirty-two, which puts the wildlife decline at 68 percent. Which figure is better representative of what's really happening?

The actual calculation of the LPI is a bit more complicated than the above example, but the gist is that the geometric mean is sensitive to large deviations, to outliers.[25] The index tells us more about extreme cases than species loss more broadly. That was the conclusion made by Canadian scientist Brian Leung and his team when they published an analysis of the LPI in *Nature*. If the 2.4 percent of species experiencing drastic population declines are removed from the calculation, the LPI actually shows a slight *increase* in wildlife populations since 1970. In fact, if it were applied to 394 species, where only one crashed by 99 percent and all the others experienced a modest 1 percent growth, the LPI would mislead us into thinking we experienced a 50 percent loss of wildlife (figure 2.1). [26] The index suggests that species across the board are under threat, when extinction risk is actually clustered in particular regions and for certain classes of living things, particularly birds and freshwater mammals in the Indo-Pacific. Many species populations are stable or even thriving.

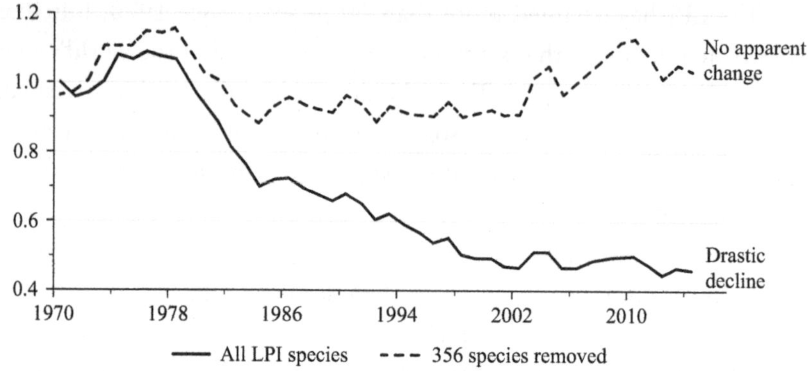

Figure 2.1
Graph of how the Living Planet Index changes based on whether all the index's species are used in the calculation or 356 of the most strongly declining populations are removed.

Leung's group concluded that metrics such as the LPI "distort the full picture" on global biodiversity. And they saw their own findings as helpful for conservation efforts. Using a measure less biased toward outliers could counter fatalistic attitudes to "provide a reason to hope that our actions [to address biodiversity declines] can make a difference."[27]

The paper sparked controversy.[28] Many of the criticisms focused on the methodology and assumptions made, such as whether the data were really representative of global trends. Simply by starting at 1970, any analysis will ignore earlier wildlife declines. Yet, what I found most interesting was how critics justified their skepticism. One reply from a team led by Michel Loreau was literally titled "Do not downplay biodiversity loss" and ended with the claim that "hope will come only from new perspectives and approaches to resolve the current biodiversity crisis once the seriousness of this crisis has been fully acknowledged."[29] Another contended that Leung's optimism was misplaced because we should set "biodiversity targets which aim to support a thriving and flourishing planet—rather than ones that 'successfully' maintain the earth in an already much-degraded state."[30]

We are back at the criteria problem. The LPI is an attractive way to judge biodiversity change if a person wants to convey the gravity of species extinction or to push for more radical biodiversity targets. And it reflects the sentiment, broadly held by environmentalists, that humans' presence on

Earth has left nature in a highly degraded condition. For Brian Leung and others, on the other hand, focusing on how losses are clustered helps us to better recognize small successes and to target the species classes and regions in most need of help. They worry about pessimistic narratives leading to fatalism or resignation.

For all the mathematical sophistication involved, this debate isn't really about the scientific merits of the LPI. The question is: Which environmental story should our biodiversity indexes emphasize? As Charles Lindblom warned, we risk hiding our own values behind models and quantitative calculation. All the layers of complexity and abstraction may prevent us from recognizing how those values shape the results or lead us to mistake the numbers for something scientifically objective.[31]

Should we use the LPI? There is no doubt that it draws attention to the loss of wildlife and rare species. On the other hand, were the LPI's biases to become public knowledge, trust in biodiversity science might be seriously undermined, especially if the measure were used to justify unpopular environmental policies. Like climate scenarios, biodiversity indexes are "scientific proxies for the seriousness" of the problem.[32] But we have to be careful about where that seriousness comes from. Does the science really demand it, and is it a story being demanded from the data?

CRISIS SCIENCE

"Nature is under siege," or that's at least how a scientific article on insect population declines begins.[33] To anyone who regularly reads news stories about environmental problems, this is familiar rhetoric. It also makes for dysfunctional politics.

The world's insects do seem to be struggling. A 2017 paper published by Dutch scientist Caspar Hallmann reported that insect biomass had decreased by more than 75 percent over a period of just twenty-seven years in nature protection areas all over Germany.[34] The total weight of insects caught by Hallman's team was less than a quarter of what it was decades earlier. Other articles followed, including a meta-analysis by Francisco Sánchez-Bayo and Kris Wyckhuys that claimed some 40 percent of global insect species risk extinction, pointing the finger at the intensification of agriculture.[35] Liberal applications of pesticides, removal of hedgerows between farmers' fields, and loss of wetlands appeared to be the likely suspects.

News stories warned of a coming insect Armageddon, an "insectageddon." Journalists relayed tales of fewer splats on drivers' car windshields, at least according to people's recollection. This seemed to corroborate the research results. The German government responded with an insect protection law that promised to eventually ban glyphosate (Roundup) pesticides and put restrictions on other agricultural chemicals. Hundreds of farmers drove their tractors into downtown Berlin in protest.[36]

This event would seem to demonstrate the political ideal for impending environmental crises: Politicians put aside their pretty disagreements and followed the science. A problem was identified, and government acted.

The "follow the science" ideal is well-known among political scientists.[37] It is the linear model of expertise. The model presumes that scientific consensus can or should determine policy. Experts impel action by speaking truth to power. But the linear model is built upon questionable assumptions—namely, that more science will provide factual certainty and that this certainty will lessen or eliminate political conflict.

The linear model's promise of evidence-based politics is illusory. As we saw for Vellend's research, the more we look, the more that we find we don't yet understand. We end up with more questions than answers, as different scientists bring divergent assumptions, methods, and theories to the same problem. The result is an "excess of objectivity," a mess of conflicting research results from which people can usually find enough evidence to confirm their already-existing political beliefs.[38] Acrimonious debates between opponents and supporters of nuclear energy are often full of references to divergent economic figures, scientific models, safety statistics, and other expert assessments, with factual consensus never seeming to actually arrive.[39] And even years into the COVID-19 pandemic, seemingly simple questions about the efficacy of mask-wearing and lockdowns, the power of the Moderna vaccine, and the origins of the virus eluded straightforward answers.[40]

Far from shunting politics to the side, the linear model leads to the facts themselves becoming the new political terrain.[41] For instance, the climate change debate quickly became trench warfare between believers and denialists of climate science, who each point to their own preferred experts and data from within the excess of objectivity. Readers hardly need reminding of the decades that have been spent arguing about whether climate change was *really happening* or really so severe. The effort to scientize climate politics ended up politicizing climate science, which detracted from the task of finding politically feasible actions to *lessen the potential risk* of climatic harms. The narrow emphasis on enforcing belief hindered the search for solutions.

Politics becomes hardened and evermore polarized under the linear model because opponents believe that winning the battle of facts will empower them to unilaterally decide the policy outcome. It is a false promise, for our disagreements only become more intractable. To some climate change believers, science tells us that we must bear uncomfortable sociopolitical changes. Avoiding catastrophe means immediately eliminating fossil fuels, forgoing automobiles, embracing new diets, and making other radical changes to citizens' way of life. Unsurprisingly, this results in opponents demanding nothing short of absolutely certain proof of impending disaster. I wouldn't call this "sowing doubt." The high level of scrutiny by skeptics was not an unreasonable reaction, given the incredibly high stakes involved if climate scientists were wrong. What if societies impoverished themselves in the effort to avert climatic changes that otherwise would have proven to be entirely manageable?

The same battle is being fought over biodiversity. The science increasingly appears to demand more radical action, such as a recent proposal to slash pesticide use by half across the European Union.[42] Farmers' groups and the pesticide lobby have been all too happy to highlight, in their view, the methodological limitations of the research. In turn, they cite their own preferred studies predicting higher prices and reduced food security. Environmentalists reply that we can easily feed ten billion people, if only we push citizens toward plant-based diets.[43] All sides seem content to line up their own facts, the research that supports their interests and worldviews, all the while acting as if nothing nonscientific is going on.

"Denialist" has already become an insult lobbed within biodiversity science.[44] Biologists James Russell and Tim Blackburn implicitly lump established researchers, who happen to be skeptical about the risks posed by non-native species, together with purported anti-science troublemakers.[45] They caution against interrupting an evidence-based discussion with "appeals to values or denial of the magnitude of the threat," dismissing skeptics as "rationalising defeat." Others have called on academic journal editors "to reconsider acceptance of denialist essays."[46]

As we've already seen, value-laden judgments are at the core of the very measures used to diagnose the seriousness of environmental problems. Whether biodiversity change is a global crisis or merely concerning depends on how the indices are put together and which data the measurers choose to give more weight. How we should go about feeding humanity is no less beset with thorny value assumptions and trade-offs. Although some groups do sow doubt about environmental problems in bad faith, self-appointed science defenders such as Russell and Blackburn come dangerously close to treating all disagreement as pathological. What starts as fighting denialism risks becoming an effort to suppress wrongthink, dictating who gets to speak about an issue and which storylines get preferred. Science's defenders too often turn into "the science police."[47]

When people try to make science fit the linear model, one storyline tends to rise to the top: undeniable calamity. Uncertain predictions get turned into clear and present dangers, into the kind of emergencies that require immediate and decisive governmental action. As science policy scholar Roger Pielke Jr. points out, the most extreme projections dominate the coverage of climate change. Nearly every gloom-and-doom story references a particular climate scenario, one that is called RCP 8.5 in IPCC reports. RCP 8.5 assumes that skyrocketing carbon emissions caused by a dramatic worldwide expansion of coal power would lead to warming reaching 4°C above pre-industrial levels by 2100. This already improbable climate scenario has only become more unlikely as developed nations have retired coal plants and invested in natural gas, renewables, and energy efficiency. But this hasn't stopped it from driving headlines and fueling dramatic scientific predictions, including for biodiversity.[48]

Although catastrophism might seem like a smart move, one that forces power to listen to truth, it can damage trust in science. The public may suspect that the facts are being used to prevent political debate. Such a loss of trust happened for climate research during the Climategate scandal of 2009. Leaked emails showed scientists planning to splice together different datasets to create a potentially misleading graph, one that nevertheless ensured that policymakers understood climate change as both novel and potentially catastrophic.[49]

The linear model's mythical presentation of science as pure and apolitical actually makes the institution even more fragile. It raises people's expectations for objectivity. As a result, skeptics view any sign of exaggerated confidence and downplayed uncertainties, or even simple mistakes, to be evidence that the science is being politically manipulated.

The insectageddon runs into even more spectacular complications than climate research. Comparing levels of carbon dioxide with global temperature is relatively straightforward. Insect scientists, on the other hand, must track myriad distinct species spread across the whole planet, including many with uncertain or unknown ecological function. More recent results have been anything but consistent. A team studying long-term ecological research monitoring sites in the United States found no clear trends in insect populations or biodiversity. A meta-analysis from scientists at the German Centre for Integrative Biodiversity Research, using still another set of monitoring sites, concluded that abundance was declining for insects that spend most of their lives on land but increasing for water-dwelling species.[50] It ends up that where and how scientists measure insect populations changes the results, making it difficult to draw conclusions from a single study—even from a meta-analysis.

Other scientists questioned the methods used in the original insectageddon studies. The meta-analysis from Sánchez-Bayo's team, which claimed that 40 percent of insect species were threatened, only analyzed studies that mentioned a decline. That is, they intentionally excluded research on insect populations that were growing. Moreover, the studies of German protected areas measured biomass, the weight of captured insects, not population numbers. Such a measure could be biased by starker declines in heavier insect species. And the researchers mainly captured flying species, so the results may not be representative of insects as a whole.[51] Indeed, one recent study found that insect species accustomed to warm weather and running water have expanded their range, whereas cold-weather and bog-loving species have declined.[52]

Surveying these studies, ecologist Manu Saunders concluded that the "biggest concern" regarding insect life is really "how little we know about it."[53] There is no doubt that pesticide overuse and habitat destruction have significant impacts, notes Saunders, but the subject is too complex to support a simple narrative of decline. Some insect species are thriving, and others boom and go bust in semi-regular cycles, preventing easy conclusions from short-term studies and popular anecdotes. Simply put, it's hard to separate a signal that unequivocally says "catastrophic decline" from all the noise of a chaotic nature.

When the German government pointed to a handful of studies to justify a ban on common insecticides, was that politics working? Although glyphosate and other chemicals very plausibly hurt insect populations, nobody had exactly proven that outlawing them was necessary.[54] On the one hand, we might see prohibition as a prudent precautionary step. Regardless of whether there really is a global insect crisis, some species populations are in decline.[55] Yet, on the other, a ban might turn out to have been a largely symbolic victory. It wouldn't be the first time. President Obama, for instance, decided to reject the Keystone XL oil pipeline, even though his own state department concluded that it had a negligible impact on oil consumption.[56] A recent study suggests that it is the droughts, cold snaps, and heat waves associated with climate change, not agricultural chemicals, that have driven down German insect populations, and it has found evidence of a potential population rebound since.[57] Future generations may look back on the insectageddon and see an environmental panic, a repeat of 1980s claims that pollution was driving a global "forest death."[58]

In an interview with BBC Radio, Sánchez-Bayo argued that "even if we don't have enough data to prove it [statistically], we know that this is happening. So, it's better to do it now than later, when we may have [a] more serious problem."[59] Although it is tempting to chide Sánchez-Bayo for appearing to run afoul of traditional expectations for scientists to "stick to the data," his statement makes sense from within the linear model. Trapped by the belief that only cataclysmic-sounding science can force governments to act, honesty about uncertainty starts to look like a recipe for delay. Many conservation scientists have told me that policymakers will only listen to authoritative pronouncements, not to lectures about scientific complexities. The linear model imprisons the thinking of policymakers, scientists, and journalists alike.[60] We will need to throw off its bonds if there is to be any hope of not repeating the pathologies of climate politics.

MERCHANTS OF CERTITUDE

According to some observers, we are heading into a "sixth extinction event" (figure 2.2).[61] Humanity's actions could bring about the disappearance of the majority of the world's species, rivaling the ecological destruction wrought by the world-ending asteroid that wiped out the dinosaurs. Seemingly taking a cue from climate change, biodiversity politics has increasingly taken on an existential tone. The problem is no longer framed in terms of losing awe-inspiring living beings or ecosystem functioning. Rather, it is posited as a threat to humanity's very survival.

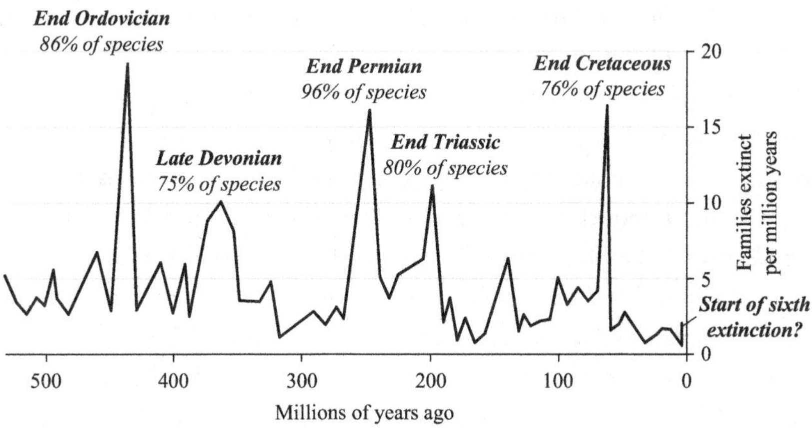

Figure 2.2
Mass extinction events throughout Earth's history. If a sixth extinction is imminent, we seem to be in its early stages.

Just like for the insectageddon, whether a sixth extinction is actually upon us is tough to say. We need to know not only how many species there really are but also the normal rate of extinction and by how much we are exceeding it. Estimates of the global number of species run from three to a hundred million. Even the commonly cited figure of 8.7 million different kinds of animals, plants, fungi, and other living things implies that the vast majority of species have yet to be discovered. Some species-estimating extrapolations seem extravagant, such as one that calculated the world's arthropod diversity after surveying a single hectare of Panamanian tree canopy. Other estimates build off of numerical trends that have emerged as we've catalogued more and more species.[62]

Many scientists calculate the current extinction rate based on the International Union for Conservation of Nature's Red List of Threatened Species. This list documents a mere 140,000 threatened species and fails to tell us much about pre-twentieth-century losses. Even then, we are incapable of counting populations at a global level. Scientists develop reasonable guesses from local and regional data. Lastly, our knowledge of the background rate of extinction comes from a fossil record that is itself woefully incomplete. It is relatively rare for living things to fossilize. An organism must die somewhere that is protected from oxygen and from things that would eat its remains. Other species, such as jellyfish, are too soft to reliably become a fossil.[63]

Estimates of how many species we have lost, are losing, and are supposed to normally lose are all complex and uncertain extrapolations. There is no doubt this makes it an exciting and dynamic area of science, but a basis for linear model science policy it is not. The scientific debate is heated. University of Hawaii scientist Robert Cowie considers the evidence to be "overwhelming" and accuses skeptics of "denial," whereas marine biologist John Briggs thinks that species losses were "exaggerated" and even that "some of them deliberately so in order to attract public attention."[64] Cowie's team uses mollusks as a proxy, as a representative sample of what is happening to life on Earth more generally, which leads him to conclude that we have lost around 10 percent of known species since 1500. Briggs, focusing on insects, estimates the modern extinction rate to be only three times larger than the normal background rate. And even then, he argues that extinctions might be balanced by the emergence of new species.[65]

Traditional assumptions that unique life forms would die out exactly in sync with ongoing habitat loss were arguably overly pessimistic. Dramatic declines in forests in Puerto Rico and El Salvador, for instance, have killed off a relatively small fraction of birds. Species have turned out to be more resilient and adaptable than early ecologists thought, at least so far.[66]

Even under a fairly pessimistic scenario, reaching a mass extinction will take at least another 240–540 years.[67] It goes without saying that a lot can change in the meantime. Even estimates of the global human population by 2100 run the gamut from more than twelve billion and rising to about nine billion and in steep decline.[68] In light of these complications, the sixth extinction starts to look a lot like the RCP 8.5, the improbable but awesomely catastrophic climate scenario cited whenever politicians or climate activists hope to sidestep messy politics. At worst, it's a kind of noble lie. The complexities and uncertainties of the issue are downplayed in favor of a dramatic and seemingly authoritative truth because it is assumed that the public will otherwise fail to understand the facts and appropriately heed expert guidance.

Seen from another perspective, such pessimism is warranted. We need only to presume that nature is inherently precarious, set on a path to destruction after a certain threshold. Ecologist Paul Ehrlich viewed each species extinction as akin to losing a rivet from a plane: At some point, the structure is compromised, and the plane crashes.[69] However, the idea that nature can tolerate only so much disruption until it is completely destroyed, that it sits on the precipice of disaster, is as much a metaphorical understanding of the environment as a scientific one. Drawing a line in the sand of warming reaching 1.5°C above preindustrial levels or of 10 percent biodiversity loss, after which comes the deluge, is unavoidably arbitrary.[70]

Although grounded in statistics and models, catastrophic predictions are ultimately storytelling. By taking nature's fragility for granted, it turns uncertain and imperfect predictions of cataclysm into a form of apocalyptic common sense. We have seen this happen already for climate change. Journalists and social media posters often depict each new drought, wildfire, hurricane, or other extreme event as heralding a catastrophic climatic future, without waiting for researchers to try to tease apart the influences of natural variability, global warming, and infrastructural failure. Devastating fires in Maui in 2023 were quickly blamed on climate change, but a strong El Niño, the neglect of agricultural lands, and poorly maintained electrical lines have emerged as contributors to the disaster.[71] Ignoring such complexities transforms impending environmental doom into a reality that can be recognized without scientists' help. Consider the 2021 editorial board of *The New York Times* pleading, "Open your eyes . . . The climate crisis is now."[72]

Politically speaking, catastrophism is a technocratic form of populism. It contrasts a commonsensical, reasonable "people" from environmental enemies whose ignorance or depravity will lead humanity to disaster. For instance, the satirical Netflix film *Don't Look Up* depicts the reality of a coming climate catastrophe as self-evident. Recognizing climatic doom is supposedly no different from perceiving a world-ending asteroid on an obvious collision course with Earth. The film's ostensible lesson is that some citizens fail to be appropriately frightened only because they are irredeemably stupid or corrupt.

The trouble is that none of our environmental crises are really so self-evident.[73] They are not like watching rising floodwaters or seeing a funnel cloud touch down. If they were, it wouldn't take millions of dollars in research to even scratch the surface regarding what is happening and what can be done about it. As philosopher Raymond Geuss put it, the truth "doesn't lie there on the street in the sun waiting to be observed by anyone who glances in its general direction."[74]

At the same time, deep disagreements persist among experts, which shows that collecting, creating, analyzing, and interpreting data aren't straightforward processes. Some think that climate change is accelerating, potentially pushing the planet over a climatic tipping point, whereas others think that the situation will be manageable. Recent research, moreover, has thrown cold water on earlier prophecies of rapidly sinking islands and abruptly collapsing ocean currents. The fact that scientists disagree proves that something must be counterintuitive or esoteric about trying to understand the significance and dynamics of environmental change.[75]

Seeing impending catastrophe as commonsensical also blinds environmental activists to important political realities. They overlook the level of trust that they are demanding from citizens. Weather extremes are normal, even if they too often cause human suffering. There is, of course, evidence that today's heat waves are incrementally hotter and longer, and that once-in-a-hundred-years storms or hurricanes now dump more rain than they should. Otherwise, life on a changing planet is relatively unremarkable, especially as improved infrastructure helps a greater portion of humanity better cope with extreme climate events.[76] Catastrophic narratives, however, ask people to doubt their own eyes when the world around them doesn't actually appear to be teetering on the precipice, or at least to put their faith in projections, models, attribution studies, and carefully curated photo montages of environmental calamity.

When people confuse complex and uncertain scientific predictions with common sense, it muddles the environmental politics. Otherwise-difficult questions get replaced with a seemingly simple one: Will we preempt disaster or not? Cataclysmic metaphors are meant to provoke an exceptional political moment, the kind of collective cooperation and individual sacrifice that is normal during a natural disaster. It is the sentiment behind Greta Thunberg's 2019 statement, "I want you to panic."[77] Roger Pielke Jr. refers to this hoped-for moment as "tornado politics," a moment in which people largely agree on what matters to them and science unproblematically clarifies what needs to be done.[78]

The implication is that people ought to make drastic and rapid changes to their lives to avert disaster and that they should do so without complaining or arguing. Tornado politics raises the political stakes to wartime-like levels. Greta Thunberg argues we can't go "back to normal"; progress demands radical, systemic changes.[79] Others contend that citizens will need to learn to see more expensive food, fewer vacations, and forgoing cars and consumer gadgets not as sacrifices but rather as the pathway to an ultimately more enriching ecological lifestyle.[80]

It is not all that surprising that such demands have fallen flat. If and when individuals do refashion their lives for the sake of the climate, they don't typically influence their environments in visible ways. Even as we witness trees being planted, birds singing, and national parks being founded, we can't really "see" a stabilizing climate or increases in the richness of global species. Rather, our individual actions have imperceptible impacts on global environmental conditions. Only multiplied a millionfold do they influence the parameters that are fed into long-term computer simulations and statistical projections. Citizens are utterly reliant on experts to tell them whether actions made any difference. It is anything but commonsensical to see lessening one's ecological footprint as akin to stacking sandbags in the face of rising floodwaters.

Yet, a disaster metaphor renders people's hesitancy for sacrifice incomprehensible. No one debates the adequacy of meteorological science at the threshold of a tornado shelter. No one derides their neighbors as Chicken Littles as water pours into their living room. Reimagining climate change as akin to a global-scale tornado makes dissent unthinkable, almost monstrous. Nearly everyone should agree when it comes to our collective survival. It is no wonder that the word "denial" gets thrown around. When it looks like human existence is at stake, it makes sense to morally equate climate skeptics or environmental optimists with Holocaust deniers.[81]

Because value disagreements stubbornly persist, global environmental problems are better described as "abortion politics."[82] Knowing more about when fetuses start feeling pain might occasionally sway someone, but generally people already made up their mind on the ethics of abortion long before high school biology class. Global environmental crises are crisscrossed with similarly value-laden questions: Is it right to eat meat? When does preventing environmental harm trump the ability of rural people to make a living? Do we trust technological change to solve the problem, or should we demand far-reaching sacrifices in the standard of living of citizens from the Global North? What level of risk should we tolerate? Is it better to bear burdens of change now or in the future?[83]

The same unavoidability of values characterizes biodiversity change. We can only evaluate it in light of what matters to us. How should we rank species diversity in relation to global poverty, nuclear proliferation, pandemic prevention, and any of the other very expensive problems that humanity faces? Our judgments on these questions invariably influence whether it makes sense to demand radical changes, such as dedicating a third or half of the earth to protected areas.

By using partial, incomplete scientific data to depict the situation as not just concerning but also apocalyptic, catastrophism waves away these broader concerns in favor of a narrow focus on survival. It offers the promise of apolitical politics. We can tell ourselves that no one would *really* lose if we were to radically transform societies to save the planet. Instead, there are only two different kinds of winners: those who recognize that they are winning and those who are too ignorant to see it.[84]

That is why the debate is so often nasty, incurious, and intolerant, why people who fail to stoically bear the necessary changes get portrayed as morally abhorrent.[85] Everyone's daily actions can be scrutinized for failing to live up to what the crisis seemingly demands. One protest sign well illustrates the tone of catastrophe politics: "Shut the fuck up about climate change if your mouth is full of meat."[86]

German journalist Sven Hillenkamp, reflecting on his years of climate movement involvement, warns that the rigidity of Fridays for Future, Extinction Rebellion, and other radical groups pushes potential allies away.[87] He laments that activists see themselves as public schoolmasters. They lecture others regarding what they must believe and what sacrifices they must make in order to realize the right environmental future. Yet, scolding results not in enlightenment but rather in mistrust. As Hillenkamp notes, many citizens are less afraid of climate catastrophe than they are of the people who promise to save us from it. Such fears are stoked when environmentalists wonder aloud whether democracy is incompatible with saving the planet and when surveys find that citizens are more willing to tolerate authoritarianism in the context of a climate crisis.[88]

Apart from simply running contrary to the basic democratic requirements of compromise and coalition building, environmental catastrophism narrows our understanding of climate progress.[89] Decarbonization has happened in the United States not via individual heroism but rather through infrastructural changes. We improved energy efficiency and gradually phased out coal in favor of natural gas and renewables. Lessening the environmental burden of meat production might not happen from enlightened lifestyle changes, but it could happen from people slowly becoming accustomed to or even preferring blending more lentils and beans into meat-based products or perhaps lab-grown meat.[90] A disaster metaphor makes us think that change has to be rapid, radical, and heroic, when the opposite strategies have done the lion's share of the work. Biodiversity conservation is unlikely to be much different from climate change in this regard.

Whatever the scientific merits of sixth extinction talk, it also carries the hope that apocalyptic scientific rhetoric can straightforwardly drive political action. Ecologist E. O. Wilson argued that scientists needed to "get [policymakers'] attention with striking and defensible facts" that a mass extinction event was already underway, and "when enough people of influence care," present them "with detailed procedures and solutions."[91] In short, it has been an effort to sow certitude about impending catastrophe, guided by the belief that the appearance of factual certainty will lead to moral clarity and political resoluteness.

Several decades of experience with climate change should disabuse us of that strategy. The only significant pieces of American climate legislation in the last decades have been the post-2008 financial crash American Recovery and Reinvestment Act and the pandemic-era Inflation Reduction Act, both of which were built upon significant compromises among multiple, often conflicting, values.[92] If we are to be more successful in collectively solving environmental problems involving biodiversity, we will have to forgo the catastrophism trump card. There may be no path forward except by directly engaging with and compromising between people's values.

BIODIVERSITY, THE VERY IDEA

Existential politics is attractive because it lets us sidestep talking about the diversity of things that people care about. Survival trumps all other values, at least so long as everyone agrees that we are truly on the precipice. If I am right in my assessment, however, the politics of impending cataclysm will not be able to deliver the policy outcomes that conservation advocates desire.

Moving toward another kind of biodiversity politics will not be so straightforward because the roots of the existential mindset are well established.

Michael Soulé argued that the idea that the "diversity of organisms is good" was a "normative postulate" for conservation, an untestable but nonetheless assumed basis for an "appropriate" attitude toward life.[93] He also saw the field as a "crisis" discipline: "Its relation to biology, particularly ecology, is analogous to that of surgery to physiology and war to political science."

When science studies scholar David Takacs interviewed some two dozen biodiversity scientists in the 1990s, he found Soulé's sentiments to be broadly shared.[94] That said, the arguments they made in favor of the concept of biodiversity were relatively diverse. Some argued that maintaining the global catalog of species differentiation mattered for human survival. Others claimed that it provided economically definable "services" such as protection from flooding or pollination of our crops.

Takacs, however, found that economic or utilitarian reasoning was usually not what personally compelled biodiversity scientists. Quite a few contended that biodiversity filled our souls, either by giving us a sense of meaning whenever we worked to protect it or by satisfying an innate drive toward biological connection. It was a holistic, almost spiritual, understanding of nature as an intrinsically valuable and harmonious balance that drove them. Botanist Hugh Iltis, for instance, saw biodiversity as valuable "in the sense that the web of life . . . all hangs together."[95] And for forest ecologist Jerry Franklin and others, this sense of "connection with other life" formed the basis for a biocentric feeling of quasi-religious fulfillment.[96]

That is why conservationists' appeals to utilitarian values often feels half-hearted, like a pretext for conversion. Stanford biologist Gretchen Daily, for instance, sees the argument for conservation as a means to safeguard important ecosystem services as just a stepping stone on the "long haul for biodiversity for its own sake."[97]

Takacs's interviews reminded me of a pair of Mormon missionaries who once interrupted me when I was shoveling my driveway. Their sales pitch was that their religion enabled a person to be "sealed" to his or her spouse, binding them together for all of eternity, unlike other Christian sects. Yet, we both really knew that the hope was that a successful conversion would turn me into a true believer, someone whose faith in Mormonism wouldn't depend on the promise of eternal matrimony.

Biodiversity scientists don't wear thin black ties and hand out religious pamphlets in people's doorways, at least not yet, but they do seem to want people to feel the earth's holy spirit. Thomas Lovejoy regularly took senators down to the Amazon, not because it was more instructive but rather because it could be a life-changing ritual. The point was to get politicians to see nature with new eyes.[98] The hope for conversion experiences is not simply a political strategy; it grows out of biodiversity's underlying holism, which extends beyond mere biology to the social sphere. Botanist Peter Raven described a "stable biosphere" as inextricably intertwined with biodiversity, "social justice," and the "human order."[99] Early preservationists pined for a broader transformation of Americans' cultural relationship with the environment into something more ecologically harmonious. Aldo Leopold once wrote, "Nothing can be done without creating a new kind of people."[100]

Seen in this context, it appears that many conservationists' real worry isn't humanity's survival but rather people's ongoing fall from ecological grace. Far too many humans continue to live in willful ignorance of the received truths of natural harmony, which threatens to deliver us all to environmental damnation. Early ecologists, such as Paul Ehrlich, explicitly saw themselves as missionaries for a new, and hopefully transformative, way of understanding and relating to nature.[101]

But it also true that developing the idea of biodiversity had become a political necessity. Biodiversity is a vaguely totalizing concept. But when compared to the similarly totalizing idea of "nature," however, it sounded more scientific, less emotionally laden. Biodiversity also addressed specific political challenges arising during the 1980s. Conservationists recognized that the emphasis on individual species, such as via the Endangered Species Act, was increasingly unpopular. The mandate to protect species from extinction nearly required forgoing a vital hydroelectric plant to preserve the migratory route of the tiny snail darter and placed considerable restrictions on logging in the Pacific Northwest in order to save the northern spotted owl.

The term "biodiversity" shifted the focus toward protecting ecosystems and habitats while still allowing conservationists to pull on citizens' heartstrings by pointing to the endangerment of charismatic megafauna, such as pandas. Instead of being forced to justify why anyone should care about a three-inch river fish, scientists could point to a more encompassing measure of ecological well-being, one that we ignore only at our own peril. This scientized the environmental debate, positioning conservationists as the main authority on otherwise thorny policy questions.

Takacs's goal when uncovering the politics of biodiversity was not to argue that researchers were pseudoscientists or charlatans. Rather, it was simply to show that the idea of biodiversity came out of a particular scientific culture, one built upon a certain story about nature. That conservation researchers developed and deployed the term with the aim to "change science, conservation, cultural habits, human values, our ideas about nature, and, ultimately, nature itself" is not really all that unique.[102] Any scientific field is going to develop a disciplinary bias. And that disciplinary bias will influence how scientists see themselves as rightfully contributing to political problems related to their areas of expertise. Genetic engineers would no doubt hope to persuade the public to take quite the opposite view of nature, one that expects to see benefits from humans steering biological change.[103]

The trouble with the original intention behind the idea of biodiversity was that it was technocratic politics. "Establishing Nature as it is understood and interpreted by scientists as the ultimate authority that human society must obey," as Ted Nordhaus and Michael Shellenberger argue, "short-circuit[s] democratic values."[104] There is no longer room for debate, only experts who need listened to. Achieving biodiversity democracy will require rethinking the standard conservation story, broadening the range of values represented within the scientific debate. Indeed, in my conversations with scientists, many of them have already been questioning this story for some time. Nonetheless, it is yet to be seen which new narratives will emerge out of the battle over the soul of conservation and which will enable us to better work through our disagreements regarding biodiversity.

BIODIVERSITY VALUES

Philosopher of science Kevin Elliott likens scientific practice to a tapestry, one woven from innumerable different threads.[105] He uses this metaphor to describe the way in which people's values shape science, impacting decisions over what to study, how to study it, how to interpret data, and what to do when we are uncertain. So tightly interwoven are data, methods, and other research decisions with value-laden judgments that it is often difficult to see where the threads of science end and those of culture begin.

My aim has been to show that biodiversity science is no different. Researchers are faced with difficult dilemmas. How much data are enough? What extrapolations are acceptable? How many population numbers are enough to make conclusions about global biodiversity? At what point can we really know that we are in a mass extinction event?

It is even unclear which words should be used when talking about biodiversity change. For instance, there is a lively debate regarding exactly how harmful it is when habitats are broken up by new development.[106] When a large forest, for instance, is split into smaller forests interspersed by grassland, farm fields, or suburbs, it used to be assumed that biodiversity goes down. However, some scientists argue that previous studies confused habitat fragmentation with *habitat loss.* That is, when they looked at biodiversity changes between big landscapes and smaller patches, scientists mistook the harmful impacts of the habitat shrinking with it being fragmented. And even then, more varied, fragmented landscapes can actually raise the level of biodiversity, although not necessarily with the same mixture of species. However, the word itself, fragmentation, suggests that something bad is happening. If critics are correct, then it would be more accurate to refer to some patchier landscapes as "heterogeneous" rather than fragmented.

Since such questions are spectacularly complex, scientists' values, both individually and collectively, are invariably influential. To try to dig down to some logically unassailable position would be a recipe for paralysis. There are real limits to scientists' collective ability to collect enough sufficiently precise data, much less make sense of mountains of information without introducing bias. Like the rest of us, scientists must partly rely on their intuition in the face of awesome complexities, especially when politics demands authoritative justifications for consequential decisions. Insofar as value-free biodiversity research is more an aspiration than reality, the belief that science can determine politics will remain an alluring chimera.[107]

Mark Vellend has taken this problem to heart. In his recent reflections, he worries about the effect of conservationists sharing too many of the same beliefs and about how most people are not well represented by his field's dominant values.[108] Vellend laments how biased some of his previous research was. For instance, he used to automatically equate non-forests with non-habitats as if they were ecological dead spaces. Nevertheless, the fact that biodiversity science is unavoidably subjective *doesn't mean* that it is "untethered to objective reality." The point is that those who claim to speak for Gaia, or to know the state of Earth's well-being, unavoidably paint their own personal values upon nature's canvas. Only after recognizing and talking openly about those values can biodiversity debates become more nuanced, less generative of public mistrust, and, one hopes, truer to what is actually happening.

For his part, Peter Kareiva has maintained a lower profile since that heated meeting at TNC. This has likely been in response to a public campaign that sought to get him fired. As with other hot-button political issues, conservation turns nasty when people believe the urgency of the problem makes disagreement intolerable. Nevertheless, his paper arguing against depicting "the vast majority of people [as] a threat to biodiversity" while "Western biologists . . . act as biodiversity's protectors" continues to be influential.[109]

Alternative stories do get told, which embrace novelty, see positives to anthropogenic changes, are sensitive to economic development, and maintain hope for incremental progress.[110] But they swim against a current of ideas that has already cut deep channels within our culture. When I read articles in the media or listen in on conservation biologists talking at conferences, traditional biodiversity sentiments seem alive and well. I hear echoes of E. O. Wilson, who saw opposition to his preferred environmental policies as "diverted and distracted by ideology," while viewing his own analyses as value-free and rooted in objective science.[111] It isn't hard to find agreement with Reed Noss's view that protecting biodiversity isn't just a good idea but also a "moral imperative."[112]

Moreover, a relatively dismal view of humanity remains common throughout the environmental movement. Horror over the magnitude of human influence on the planet combines with an ever-present cultural attraction toward apocalyptic thinking. Many see humans as akin to the dinosaurs at the end of the Cretaceous period, except that humanity is not just the doomed species but also the world-ending asteroid itself.[113]

The following chapters dive more deeply into different facets of the biodiversity problem. My aim is to show that when we debate the harms of alien species or the promise of rewilding ecosystems, we are talk not just about data and facts but also about different ways of valuing nature and of imagining humanity's place within it. Again, I won't pretend to know who is right. Rather, my focus will be on laying the groundwork for a more pluralistic biodiversity politics. Only by taking different ways of understanding nature seriously can we open up possibilities for actually resolving environmental disputes in productive and democratic ways.[114] Decades of science-based conversion attempts and calls for environmental heroism have only left our societies more polarized.[115]

If biodiversity is going to be different than climate change, it will only be by accepting some uncomfortable realities. Many people will never be convinced of the sixth extinction or of a coming insectageddon. They will refuse to transform society or sacrifice their standard of living in order to appease indexes and models claimed to represent the earth's ecological wellness. And they will have good reasons for doing so. Declarations of planetary crisis will never strike many people as commonsensical.

We will need to move beyond the polarized politics of believers and deniers, to get away from trying to turn environmental science into a form of spiritual guidance and scientific experts into moral lodestars. As climate policy scholar Mike Hulme argues, "There is no trick that will force a convergence of human values."[116] Environmental politics needs broader and more diverse coalitions. This means figuring out how something like biodiversity can matter to citizens on their own terms rather than insisting upon the values embraced by the self-appointed guardians of nature. Reconciling biodiversity with democracy demands it.

3 INVASION

It was cocaine traffickers who first brought the hippopotamus to Colombia. Now, conservation scientists want it gone.

Born in an Antioquian village in 1949, Pablo Escobar came of age in the sprawling metropolis of Medellín. Known today as one of the most infamous drug traffickers of the twentieth century, Escobar's life in crime began modestly. As a youth, he sold forged high school diplomas and stole oranges. Eventually, he moved on to start a business in misappropriated gravestones. However, it wasn't until he discovered the cocaine trade that Escobar's career really took off.[1]

Getting into the smuggling business was pure serendipity. During a two-month stint in the La Ladera prison for auto theft, Escobar met Alberto Prieto, a legendary smuggler known as "the Godfather." Prieto gave him a job guarding his truck caravans. It was during this time that Escobar learned of Colombia's "kitchens," whose workers imported coca paste from Ecuador and cooked it into cocaine.

With his cousin Gustavo, Escobar eventually devised a system to import increasingly larger amounts of coca paste. Day laborers would deliver it in hollowed-out pieces of wood to countryside kitchens. Escobar soon recognized that only vertical integration would enable his business to grow. In the late 1970s, cocaine became a fashionable drug among affluent people in North America. Law enforcement was slow to react, and port inspectors and airport agents could be fooled by suitcases with false bottoms. Eventually, the Medellín Cartel, headed by Escobar and shipping cocaine by the ton, controlled the drug trade in Florida.

Escobar was a wildlife lover. As he exported cocaine to the United States, he simultaneously imported exotic animals from across the globe. His sprawling Nápoles estate in the Magdalena Medio region featured a collection of wildlife, including elephants, zebras, giraffes, camels, kangaroos, tigers, lions, rhinoceroses, and, more importantly to our story, hippopotamuses (figure 3.1). [2]

The Colombian government declared war on Escobar after he ordered the assassination of the Minister of Justice, something Escobar did in retaliation for the relentless investigations of his cartel's drug trafficking. A cycle of arrests, extradition attempts, and violent reprisals continued for years, a situation further complicated by an ongoing battle with a competing cartel and threats from a vigilante group formed by his enemies.

After Pablo Escobar died in a shoot-out with police, state agents were tasked with removing the exotic wildlife from his estate. They inexplicably left the hippopotamuses to their own devices. The animals are notoriously difficult and aggressive, taking more human lives in Africa than big cats. A quick Internet search uncovers videos of hippos easily shrugging off lion attacks and occasionally turning around to deliver a fatal bite on an

Figure 3.1
Hippopotamuses swimming in a muddy pond near Pablo Escobar's former home.

overconfident big cat. Perhaps agents believed Escobar's imports would not survive unassisted, living so far from their natural habitat, but the animals thrived in the Magdalena River system.[3]

By the late 2000s, some local residents considered the hippopotamuses to be a nuisance. The regional environment ministry hired professional hunters after receiving complaints that one ornery male, called Pepe, had knocked down fences and destroyed crops. Pepe died by a bullet and at the federal government's behest, just like his former owner. Public outcry over his death was swift and intense.[4] In the following decade, the Colombian government hesitated to cull the ever-larger hippo herd while demands for action from conservation scientists only grew louder.

Colombia's cocaine hippos are more than just a fascinating sidenote to a true crime story. They constitute a natural biodiversity experiment: the return of giant animals to a South American river system thousands of years after their disappearance. The case evokes many of the same kinds of dilemmas faced elsewhere in conservation. What does it mean to be a "native" species? How endangered are environments by alien species "invasion"? Is it humanity's role to keep ecosystems as native as possible?

Many of us tend to think of non-native organisms as always bad. The term "alien" species evokes images of stowaway rats decimating Pacific islands or overly confident land managers introducing foreign predators to control pests, only to be overwhelmed by them. However, the answer to the question concerning non-native species and whether hippopotamuses should remain in the Magdalena River basin is not as scientifically clear-cut as one would think. Here as much as elsewhere, we face a thornier and much more subjective and political conundrums: On which species' behalf should we act? Which nature are we trying to protect?

A HISTORY OF INVASION

"An unprecedented biological invasion." A "critical threat to biodiversity." Reading the scientific literature, it would seem like the Colombian hippopotamus must be an open-and-shut case of a harmful alien species. Indeed, two professors argued in a letter to the journal *Frontiers in Ecology and the*

Environment that it was only because of "science denialism" that a culling program had not yet been instituted. They portrayed the existence of hippopotamuses frolicking, if you could call it that, in Colombia's rivers as yet another case where public irrationality is privileged over authoritative research. We need only to follow the science to get things right.[5]

There is good reason to worry about what an African megaherbivore might do to a South American landscape. The majority of the continent's very large animals, or megafauna, including potential hippopotamus predators, died out some twelve thousand years ago.[6] In the language of ecologists, Colombia's ecosystems may now be "naive" to the activities of a two-ton herbivore, a naivety further exacerbated by the fact that the hippopotamus is a newcomer. Colombian plants and animals may struggle to adapt to the giant animal's impacts on the environment. Native species might even go extinct.

This is what it means for a species to be invasive. When humans introduce animals, plants, insects, or microorganisms outside of their traditional geographic range, those species sometimes manage to sustain a large population that produces "demonstrable environmental or social-economic impact."[7] The fear is that these car-sized animals enjoy unmitigated advantages in their new environment, allowing them to dominate it at the expense of humans and other species.

Most of us can name an invasive alien species that has wreaked havoc on a native landscape. Among the most notorious are Australia's cane toads. In the early 1900s, cane grubs were ravaging the sugar fields of Queensland. The cane toad looked to be a highly effective form of biological pest control, having already proven themselves in the sugar fields of Puerto Rico, Hawaii, and the Philippines. Farmers were optimistic, with one jubilantly declaring, "We've got these bloody grubs by the balls this time. And we'll go on to bigger and brighter things." But the Aussies' grip on the grubs' gonads wasn't as firm as it seemed. Contrary to experiments that appeared to confirm the toads' usefulness, their introduction to Queensland farms in 1935 turned out to be an abject failure. Although the toads ate cane grubs in the relatively rainy climate of Hawaii, they shunned Australia's drier cane fields.[8]

The cane toad's second-rate grub-eating soon became the least of Australians' worries. Native to the Americas, the toads have fewer natural predators in Australia. They excrete a toxin from a gland on their shoulders, which has been known to kill numerous native species that try to eat them, along with domestic cats and dogs. Yet, some rat, snake, turtle, crocodile, and bird species are known to consume cane toad without ill effect, so it is not as if the species is without any threats at all. Nevertheless, these new arrivals feasted on snails, mollusks, and other species previously ignored by native frogs, and the rotting corpses of naive toad predators soon dotted the landscape.[9]

The impacts of the cane-toad invasion were not limited to the environment. Some Aussies claimed that drinking a solution made from boiled cane toads triggers a vivid psychedelic experience. It remains unclear how prevalent the practice really is. Some experts warn that the toxin will sooner end a person's life than give them hallucinations.[10]

Humans have assisted biological invasions of one form or another for thousands of years. Ancient Central Pacific seafarers are estimated to have precipitated the demise of more than half of the bird species on the islands they colonized, partly by introducing rats, pigs, and other novel species.[11] People have a long history of unintentionally rejigging ecosystems.

Nevertheless, alien plant and animal populations have significantly increased since the mid-1850s. Some of these introductions were intentional, such as importing ornamental plants for a garden or releasing a predator for pesky agricultural pests. But other cases were the accidental by-products of a simultaneously globalizing and warming world. Species stowed away on ships and people's shoes or migrated in the face of a changing climate.[12] These invasions cost us $25 billion per year, according to one estimate, with the money being spent on pest control or repairing damaged infrastructure.[13]

The introduction of non-native species has come to represent the worst of humans' environmental meddling. News reports on breaking studies warn readers that increases in alien species will lead to dramatic and permanent damage to ecosystems as environments are pushed beyond "biological tipping points."[14] Although such warnings certainly sound alarming, for many cases, the evidence is stubbornly ambiguous. The more one researches the story around alien species, the more complicated the picture becomes.

THE METAPHOR IS THE (SCIENTIFIC) MESSAGE

The poison is dropped by helicopter. It is hidden in food. When the target consumes it, they are fine for a while—otherwise they might figure out that something was amiss. They start to stumble as they lose the ability to coordinate their body. Their eyes take on a sunken, glazed-over appearance. Then, the vomiting starts. It takes anywhere from four to eighteen hours for death to finally arrive. In the meantime, the target lies on the ground, gripped by spasms and tremors. If the poison doesn't do them in, traps serve as a backup. The devices are spring-loaded, designed to fracture the skull in a violent snap.[15] Although this reads like the depiction of a dictator trying to quash a rural insurgency, it is how many countries control non-native species.

When I ask biologists who is responsible for our hawkish attitude toward non-native species, they usually name Charles Elton. He was a small, balding, and unassuming biologist who trained at Oxford. Elton, nevertheless, had a flair for the dramatic. He would brag to colleagues about how he used to box and that he had even once knocked out another man. But the boast would be delivered in a near whisper and from a man wearing a sweater vest.[16]

Metaphors, including warlike ones, have been common in biology for some time, filling the pages of Charles Darwin's *On the Origin of Species*. But it was Elton's 1958 book, *The Ecology of Invasions by Animals and Plants*, that many credit with cementing the term "invasion" in scientists' vocabulary.

Elton's book is awash with literary allusion. Introduced species are "invaders" who cause "ecological explosions." Notably, he starts off by telling the story of the bubonic plague, a bacterial infection spread by the fleas that sheltered on rodents. By taking an epidemic disease as his model, he sets up alien species to be inherently dangerous foreign contaminants in an otherwise healthy or "cosmopolitan" ecosystem. He warns the reader that they are witnesses to "one of the great convulsions in the world's fauna and flora," quoting Sir Arthur Conan Doyle to portray the war against biological invasion as "one of the typical decisive battles of history—the battles which have determined the fate of the world."[17]

You don't need a PhD in science studies to recognize that Elton's take on non-native species is as much storytelling as science. Environmental scientist

Brendon Larson warns that the underlying militaristic metaphor primes us to misunderstand the impact of introduced species and blinds us to other ways to dealing with them. We come to imagine our task as ensuring that ecological "good will triumph over evil" in order to "return ecosystems to a pristine state."[18] Indeed, the Convention on Biological Diversity directs signatories to eradicate or control alien species that have a high potential for environmental spread. And biodiversity indicators usually exclude non-native species. Even relatively benign introductions typically don't count as part of the natural environment.[19] War against ecological newcomers is built into international treaties and scientific measures. It is the world that Charles Elton would have wanted.

The irony is that environmental war metaphors were earlier popularized by the chemical industry, a traditional enemy of conservationists. The post–World War II explosion of chemical remedies for pests and weeds was heralded by scientists and farmers as a godsend, providing people with a much-needed weapon against the insect and plant species that threatened human survival. Memories of entire fields of crops lost to pests was fresh in the minds of many farmers and agronomists.[20]

The depiction of nature as a battle of survival between humans and the pestilent swarms soon gave way to another story: nature as a fragile web of interdependent relationships.[21] Nevertheless, the notion of undesirable species lived on, now being used to stigmatize the animals, insects, and plants that appeared to upset the delicate balance that mid-twentieth-century ecologists believed they saw in nature.

Many scientists and journalists now question the battle against species deemed to be out of place. Control requires considerable expense and for the invading species to suffer.[22] The South Pacific islands have been repeatedly dusted with poisons. Feral goats in the Galápagos Islands have been shot from helicopters. Rabbits have been deliberately infected with viruses. However, for land areas bigger than islands, the situation looks more like trench warfare. Journalist Emma Marris cites efforts on New Zealand, where poisoning 95 percent of the possums and all the rats in a region buys cherished flightless birds merely "one or two good breeding seasons." Protecting nativity means we "must shoot, poison, trap, fence, and watch, forever watch, lest the excluded species find their way back in."[23]

The case for this kind of biological warfare is complicated by two ongoing debates. First is the extent to which nativity is a coherent biological concept. The second involves the degree to which non-native species are typically, predictably, and irreversibly harmful to other species and the environment.

OUT OF PLACE

Where do camels belong? This seemingly simple question is the title of botanist Ken Thompson's polemic against the idea of alien species. The question alludes to the fact that camel-like species first emerged in the North American plains, even if most of us think of camels as domesticated, Middle Eastern pack animals. Camelids enjoy their greatest diversity in South America as llamas, alpacas, and other related species. And the biggest wild and self-sustaining, albeit imported, population can be found in Australia. Although extinct in their traditional homeland, camelids were almost accidentally restored in the 1850s. The United States imported camels as beasts of burden during westward expansion, an experiment that was ended by the protests of (you guessed it) the mule lobby and the Civil War. In any case, the example shows that it is sometimes not so easy to answer the question "Where is this species *supposed* to exist?"[24]

In order to label a species as either native or alien, we must choose a moment in time, a certain snapshot of an ecosystem, that defines nativity. Thompson, among others, highlights the arbitrariness of this practice. In the Americas, 1492 is often set as the baseline for deciding what belongs, whereas in the United Kingdom, native species are whatever one finds in Jane Austen novels. But some "native" species were merely introduced long ago or liked well enough for most people to forget their origins. The Spanish have been hard at work controlling the invasion of North American crayfish, even though the threatened white-claw crayfish was most likely itself introduced in the eighteenth century. Borneo's endangered pygmy elephant, despite being the target of conservation efforts, may have arrived only a few centuries ago.[25]

Species have always moved around. As biologist Chris Thomas recounts, the ground in a typical United Kingdom back garden has seen a range of climates: grassland, forest, steppe, and so on.[26] Different animals, plants, and insects arrived and disappeared alongside climatic changes. Furthermore, species have colonized new habitats, or were alternatively outcompeted, when continents have collided. After the two Americas connected some three million years ago, northern species rushed across the Panamanian isthmus, soon composing half of all the mammals in South America.[27] Invasion and extinction happen without human interference.

Extinctions, however, offend our sensibilities. They snuff out genetically unique species. Federal officials, acting in the name of conservation, plan to shoot a half million barred owls in the Pacific Northwest. The birds have been sentenced to death for the crime of encroaching on the habitat of the less common and more charismatic spotted owl, a species that has continued to decline despite decades of protections. The two species are as much cousins as they are competitors. Whether their fertile offspring symbolizes environmental contamination or the production of a more resilient hybrid depends on whom one asks. The fact that the genetic uniqueness of the spotted owl forms the legal basis for logging restrictions no doubt complicates the situation.[28]

The greatest ire is saved for the species out of place because of human meddling, for organisms that bridge distances and arrive in far greater numbers than natural migration would normally permit. Our outrage at the spread of cane toads in Australia is amplified by the fact that we brought them. Bird-decimating rats stow away on our ships. People introduced a rampant fungal infection now wiping out amphibians across the globe. According to one theory, chytrid fungus piggybacked on the African clawed frogs imported for use in an early form of pregnancy testing, animals that some hospitals irresponsibly released into the wild when they were no longer needed.[29]

The idea of nativity sets strict conditions for belonging, all but ignoring whether an alien species could ever be "naturalized." Consider the North American black cherry tree. When it was brought to Europe, the species' ability to spread quickly earned it a reputation as a "forest pest." Nonetheless, scientists have found that many native insect species have adjusted to eat the newcomer's leaves, and it is now fed upon no less than native bird cherry. An introduced species can evolutionarily integrate into its new ecosystem and "may eventually reduce its invasive character to the point where it attains the status of non-harm [and] naturalized."[30] The question, of course, is whether people find the speed at which this happens to be acceptable.

Hippopotamuses seem to only belong in Sub-Saharan Africa, although they waddled through a few European rivers prior to the last Ice Age. Today, they are found in isolated pockets along African rivers and are vulnerable to extinction. Their presence in Colombia is clearly artificial. Even though they are surprisingly able swimmers, no hippopotamus has been spotted trying to migrate across the Atlantic. However, the case might not be so black and white. One set of researchers has suggested that the hippopotamus is a passable substitute for the now extinct *Trigonodops lopesi* (figure 3.2),[31] also a semi-aquatic megaherbivore.[32] It is not a case of direct descent, such as if we introduced wild Australian camels to North America. Nevertheless, a

Figure 3.2
Artistic depiction by Jorge Blanco of a *Toxodon*, a South American genus that some scientists believe the long-extinct but hippo-like *Trigonodops lopesi* would have belonged to.

hippo-like animal was native to South America. Perhaps that is not enough to earn the animal a Colombian resident visa, but it makes it harder to dismiss the species as a totally alien invader.

REENGINEERED LANDSCAPES

Beavers were introduced to Argentina to support a new pelt industry. Instead, the musky and tenacious animals have chewed up the Patagonian landscape. Beavers dam up streams and gnaw down trees in Canada too, but South American beech species are less tolerant of flooded soils and don't like being cut. In Argentina, beavers' insatiable desire to reengineer their environments arguably becomes a bug, not a feature. Large sections of Patagonian forest now appear as if they had been clear-cut by loggers.[33]

When environmental scientists look upon such a landscape, they typically describe it as having been "degraded."[34] However, when reading through the research literature, that concept is less robust than one might expect. Newly arriving plant species, for instance, can alter the properties of the land, such as the mixture of fungal and microbial species in the soil. Sometimes newcomers crowd out native species or make space for other non-natives. They can cause the soil to dry out faster, increase the likelihood of fires, and amplify erosion rates.[35]

It is tricky to characterize these changes as bad in a nonsubjective way. Although we can explain why we might not like the impacts, giving environmental harm an objective definition is a different matter. A lot of policy simply sidesteps the thorniness of the problem. For instance, the US National Invasive Species Council vaguely defines environmental harm as "alterations to plant and animal communities or to ecological processes that native species and other desirable plants and animals and humans depend on for survival."[36] When it comes to alien species, the presumption is often that any post-introduction changes will be bad.

The impact of the hippopotamus has not been as stark as that of Patagonian beavers. The environment looks little different from before, only there are three-thousand-pound giants chilling in the rivers by day and grazing nearby fields at night. Why should we care if hippopotamuses end up someplace where they *shouldn't* be?

The problem is that hippopotamuses shit. A lot. An East African hippo's annual deposit to his local river might weigh in at three quarters of a ton, roughly the same as six full-sized refrigerators.[37] While some worry about hippopotamuses crowding manatees and capybaras out of their aquatic niche, most scientific papers focus on the risk of "eutrophication" produced by the animal's penchant for river pooping. The ensuing brown tsunami increases water nutrient levels, causing algal blooms that suck up the available oxygen, in turn suffocating fish.[38]

Is this ecological degradation or just a situation where sometimes fish lose and hippopotamuses, algae, and species that feed on decaying fish win? Likewise, should the beaver's disruptions earn it a death sentence? Picking environmental winners and losers feels like a dismal form of environmental calculus. How many rescued beech trees justify culling a hundred thousand beaver? Do the harms of non-native grasses growing in beaver-engineered meadows outweigh the apparent benefits that beavers provide in Tierra del Fuego—namely, providing habitat for introduced brown trout, native puye fish, and Magellanic woodpeckers?[39]

An alternative approach is to view the changes in terms of environmental complexity and resilience. Disruptions are a normal part of how an ecosystem functions. So, the question is not whether an introduced species has environmental impacts but rather whether the ecosystem is simplified by the newcomers' influence on it. Just as a field of genetically identical corn plants risks being decimated by a disease or pest, nondiverse ecosystems are less resilient to future disruptions. Yet, as we saw in chapter 2, new species can also add diversity over time, not just take it away. The wickedly hard question remains: Will an ecosystem impacted by an invasive species necessarily continue to degrade or will it eventually bounce back and adapt, perhaps becoming even more biodiverse and resilient?

The trouble is that the answer to that question may come too late. Harmful species introductions can be really, really bad. The textbook example is the brown tree snake. Introduced to Guam during World War II, the snake snuffed out nine out of eleven local bird species and likely put the final nail in the coffin for native bats. The fauna of Guam had no experience with large, poisonous snakes. The population density of brown tree snakes has reached thirteen thousand per square mile in some places, and the species grows several feet longer on Guam than in its native habitat.

It seems unlikely that the hippopotamuses' presence will rival that of the brown tree snake. Most extinctions from introduced species happen on islands, where populations are small, local species have no escape, and natives have usually evolved to a fascinatingly unique but sometimes defenseless state. Even so, we humans often find it discomforting to witness other animals engineering their environments, especially when the aesthetics offend us. However, as science journalist Ed Yong put it, the natural state of many of Africa's rivers may not have been "a babbling brook of clear water" but rather (at least sometimes) a "world of dead bodies, putrefying [hippo] poop, and the occasional wave of suffocation."[40] Periodic fish kills from eutrophication are a normal part of African river life.[41] How can we be sure such a situation is any more undesirable in Colombia's rivers?

SCAPEGOATS

The roar of woodchippers is overwhelming. The smell of diesel fuel fills the air. This is what decades of control of tamarisk trees or salt cedars in the American Southwest has come to. Making progress against these invaders from the east literally requires heavy machinery.

Living in New Mexico, I have heard a variety of commonsensical explanations for salt-cedar eradication. The shrubby trees were thoughtlessly introduced to control erosion but soon became "water thieves," sucking the local landscape dry and leaving behind a salt-encrusted earth. Without salt cedars, native cottonwoods would instead line the shores of the Rio Grande and other Southwestern rivers.

But that story is wrong, according to biodiversity researcher Matthew Chew. Tamarisk trees don't actually use more water than native plants. Their hardiness and success in the arid environment just made them an easy target for Western residents looking for someone to blame, especially as droughts undermined already-marginal desert agriculture.[42]

Even worse, tamarisk seems to outcompete native cottonwood trees mainly because our dams prevent the periodic floods that cottonwoods need to reproduce. Rationalizing the river for human use has left the shoreline saltier than before and lowered the water table, both of which work to the salt cedar's advantage.[43] The problem isn't that the tamarisk is a superweed wreaking havoc. Rather, it is that people have created a landscape that favors the species. Portraying salt cedars as environmental monsters obscures humans' own role in bringing about their spread.[44]

This should ring true to backyard gardeners. By pulling weeds and trying to grow a grass monoculture at my New Mexico home, I invariably create an ecological vacuum that crafty, generalist plant species are happy to move into. Even six-millimeter plastic sheeting and several inches of gravel can't stop them. I often feel like Bill Murray's character in the movie *Caddy Shack*, whose efforts to rid a golf course of a pesky gopher are increasingly extreme (eventually involving dynamite) but no more successful.

In any case, science journalist Fred Pearce believes that other invasive species have been unfairly blamed. In his book, *The New Wild*, Pearce contends that the invasion of water hyacinth and Nile perch in Lake Victoria, Africa's biggest body of fresh water, was driven primarily by high levels of pollution.[45] If Pearce is right, the demise of countless cichlid fish species was less due to the perch than to humans creating the conditions for the newcomer to thrive. Like any good scientific question, what exactly happened remains hotly debated.[46]

Similarly maligned are the wild burros that now roam Death Valley, California. The donkeys are considered a destructive species by the National Park Service, charged with the typical laundry list of invasive impacts: sucking up scare water, overgrazing, digging holes in the ground, and pushing out natives. Yet, biologist Erick Lundgren argues that the burros' digging actually provides ideal sites for cottonwood tree seeds to germinate and exposes groundwater for other species. Their grazing activity then keeps local springs from being choked out with vegetation. More importantly, Lundgren maintains, the species' negative impacts may have less to do with the burros' activities per se than with the fact that mountain lions are routinely killed to protect livestock and native bighorn sheep. Donkey populations may be booming only because, by eliminating predators, we have handed them a big ecological advantage.[47]

The monstering of the Colombian hippo may be similarly unfair. Eutrophication is not caused by hippopotamuses alone. The Magdalena River valley is a modified landscape, dotted with farms whose runoff also increases nutrient levels. In fact, the ponds in which Colombian hippos currently wallow are themselves man-made.[48] This isn't a pristine ecosystem only now being imperiled by an invasion. Human activities both add to the eutrophication problem and have aided the hippopotamus' spread.

FRIENDS OR ENEMIES?

Carp have run rampant in America's Midwestern rivers since their accidental introduction in the late twentieth century. In many areas, they are now the dominant fish species, sometimes injuring boaters as they jump from the water. Hundreds of millions of dollars have been spent to prevent the uncertain consequences of the species migrating to Lake Michigan, including installing electric barriers to "teach" the fish to stay put. Officials have even gone so far as to dump six tons of fish poison in the waterways, the biological equivalent of scorched-earth warfare, to avert what they see as an ecological catastrophe in the making.[49]

The research is not altogether clear that the carp is as harmful as feared. The species' main mistake may have been thriving on the wrong continent. Carp is farmed at considerable effort and expense in Asia. One wonders if the situation wouldn't be perceived as so dire had Han Chinese rather than Europeans colonized the Mississippi River valley. Only the latter shun the fish's bonier fillets. The carp is controversial partly because it impacts species, such as crappie and large-mouth bass, that Americans prefer to eat. Local agencies catch up to a million pounds of the fish each year just to turn the harvest into fertilizer or dump it in landfills.[50]

Controlling alien species is as much about humans' interests as environmental well-being. For example, zebra mussels clog utility water pipes and crowd out mussels that we would otherwise catch. Japanese knotweed is feared for its reputation of growing into and expanding cracks in concrete foundations and brickwork.[51] Alien plants can crowd out forage for livestock. Imported pests wipe out acres of trees that could be otherwise logged. Species that threaten the things people want are more likely to get seen as ecosystem enemies.

Some invasive species, however, come to be highly valued, even prized. Feral pigs are culturally important to Native Hawaiians, despite their role in changing the landscape. Non-native wild horses are a protected species in the American Southwest. The Nile perch's booming population in Lake Victoria supports fishermen. Across the world, alien species are important sources of food, fuel, and medicine, or aid in erosion and pest control.[52]

The situation is only further complicated by evidence that newcomers can benefit native species. The endangered southwestern willow flycatcher, for instance, is happy to nest in tamarisk trees.[53] Native species advocates face a dilemma when it comes to grinding up salt cedars. Although the trees themselves came from elsewhere, they make perfectly good homes for local birds in a watershed forever changed by man-made dams.

Tamarisk trees aren't the only alien species that can be friends with natives. Imported honeybees, rats, and birds can take over some of the job of native pollinators. Non-native trees can help fickle native species move back into abandoned pastures. Otherwise-invasive mussels and oysters clean up waters polluted by human activity. One particularly interesting example is a barrel-shaped animal called *Pyura praeputialis*, which has colonized the rocky shores of Chile. Although their population numbers would suggest that an invasion has occurred, they reengineer the environment in a way that enhances the local ecosystem. The habitats with *P. praeputialis* boast double the biodiversity of those without. The relationship between alien and native species isn't necessarily zero-sum.[54]

Others argue that non-native species shouldn't be thought of as invasive but rather as successful. Biologist Chris Thomas devotes a lot of pages in his book, *Inheritors of the Earth*, to the lowly house sparrow. Often derided by birders as alien and boring, Thomas, in contrast, celebrates the species for its knack for coexisting with and even thriving among people. They are fit for the Anthropocene. And house sparrows are not the only ones. Raccoons, white-tailed deer, squirrels, coyotes, and foxes seem to readily adapt to human presence, at least up to a certain point.[55] Even in highly modified environments, species can find a niche for themselves.

This successfulness strikes many conservationists as ill-gotten, especially if it happens in the countryside rather than in a suburban neighborhood. Monterey pines barely hang on to life in their native central California home, earning the tree a spot on the International Union for Conservation of Nature's Red List of Threatened Species. Strangely enough, Monterrey pine grows so well in countries such as New Zealand and Chile that many lumber farms plant the tree almost exclusively. But that same adaptability earns the species a reputation as an invasion threat. Yet, it seems at least a little perverse that humans break their backs to save a seemingly fickle and sickly tree in California while simultaneously pulling up any tenacious saplings that escape plantations in New Zealand.[56]

So, are hippopotamuses friends or enemies? Although early twentieth-century Americans toyed with the idea of raising the animal for meat in Louisiana, river-horse steaks are unlikely to make an appearance in Colombian restaurants. That said, many locals see the species as a potential tourist attraction.[57] Moreover, hippopotamuses' prodigious waste production may not be an ecological bug but rather a feature. This spreading of nutrients can reengineer local environments. Even today's relatively low population of seabirds and migratory fish species, such as salmon that travel upstream to breed, move more than a hundred million pounds of phosphorus from the oceans to land each year. Hippopotamuses have even been described as essential "conveyor belts" for moving nutrients from African grasslands to aquatic ecosystems. And some scientists speculate that this conveyer belt helps feed manatees.[58] The future of the Colombian hippo will depend on where we draw the fuzzy line between a successful or useful species and an invasive one.

WHICH SPECIES? WHAT HARM?

Many scientists, such as ecologist Daniel Simberloff, argue that we ought to continue eradicating invaders.[59] His 2013 book, *Invasive Species*, exhaustively catalogues all the ways that introduced species can alter ecosystems and reduce native populations. Very much the successor to Charles Elton, Simberloff celebrates the elimination of alien black-striped mussel in Australia and of rats, pigs, and goats on different islands. He finds solace in these cases. He sees the potential to return landscapes to something more like their previous states.

Yet, like elsewhere in conservation biology, a lack of evidence undermines our ability to make sweeping statements about the invasiveness of alien species per se. One study estimated that only 10 percent of the species introduced to Europe had even been documented.[60]

Despite the unrelenting uncertainties, there are rules of thumb. The classic "tens rule" asserts that only 10 percent of non-native species can survive in their new environment, only 10 percent of those establish a stable population, and only 10 percent of those species cause ecosystem harms. A recent paper casts doubt on the validity of this rule, arguing that likelihood at each step is not 10 percent but anywhere from 25 to 50 percent.[61] Even with this revision, it is still only a small minority of non-native species that cause problems. Indeed, according to a 2021 study, one implicating non-native species in 20 percent of known extinctions, most of the blame can be placed on a few usual suspects—namely, rodents, cats, mongoose, Nile perch, and chytrid fungus.[62]

Where introductions don't result in extinctions but rather more moderate environmental disruptions, we are beset with a thorny complication: What mixture of species is *supposed* to be in any given place?

A recent *Wired* article warns about an unexpected new invader of the American plains: trees.[63] To a lay person, this is understandably perplexing. After years of hearing that deforestation and ever hotter wildfires harm wildlife, the idea that trees undermine biodiversity certainly feels odd. Clearly, less grassland is bad for ranchers and species that prefer open spaces, and human fire control no doubt tips the ecological scales in trees' favor. But grasslands and forests have advanced and receded for millennia. Why should it matter to the planet if there are "too many" saplings emerging on the plains?

An accidental experiment in Kenya illustrates the underlying dilemma. Cattle and wild ungulates have roamed the country's Laikipia plateau for more than three thousand years. Scientists protected several research plots from these herbivores. Protection resulted in a tripling of woody plant species richness after a mere sixteen years. But when elephants later broke into the enclosures for a snack, species richness declined, with rarer native plants tending to suffer more. If the local diversity of native plant species is prioritized, the reintroduction of large herbivores could look a lot like an invasion. But the original variety of plants was itself artificial, brought on by humans excluding large grazers and browsers.[64]

We are, once again, stuck with the dilemma of trying to decide which historical moment defines a good ecosystem. For the Laikipia plateau, should we prefer the one with more megaherbivores or the one with more diverse plant species? Similarly, for the American plains: How do we know the emerging forest ecosystem is worse than the grassland that preceded? Species numbers alone can't answer that question.

Given that the hippopotamus might be a passable substitute for the long-dead *Trigonodops lopesi*, are the river horse's impacts actually a degradation of the Magdalena River valley? Or might the megaherbivore bringing the ecosystem back to something more like its previous state? Which artificial collection of local species is *more* natural?

To some scientists, conservation biology errs in being so backward-looking. The tendency to view an ecosystem populated by newcomers as invariably degraded blinds us to the surprising resilience, or even vibrancy, of ecosystems in the face of alien migration or introduction. Rather than see these places as irrevocably damaged, they are merely "novel."

Like the *Pyura*-invaded tidewaters of Chile, some novel ecosystems are more diverse and no less functional or productive than neighboring "intact" ones.[65] This has led some biologists to see them as alternative to never-ending intensive management of ever-vulnerable pristine parks. Critics, however, believe that ongoing invasion will eventually lead to ecological simplification, the ecosystem equivalent of "eating at McDonalds."[66] Nonetheless, as for introduced species more generally, the balance of harms and benefits for novel ecosystems can only be evaluated on a case-by-case basis. Biologist Joseph Mascaro and others have found that forests in Puerto Rico seem to have taken on immigrants without any long-term harm. In fact, they seem more productive, and nutrients cycle through them more quickly. In contrast, natives are more clearly being crowded out by aliens in Hawaii. Although ecosystem functionality seems unchanged, some parts of the islands are now more prone to wildfires.[67]

Although the hippopotamus is disruptive to the preexisting ecological relationships in the Magdalena River valley, it is possible that the ecosystem may eventually settle into a more vibrant state. Exploring that possibility, however, leaves us adrift in a sea of risky uncertainties. An ecosystem's historical baseline offers us a comforting reference point, without which we are forced to more explicitly define ecosystem health. That question doesn't have a straightforward scientific answer because any reply to it rests upon beliefs about how the natural world functions and the proper place of humans within it.

WHICH NATURE?

When people face uncertainty, they invariably fall back on truisms. "Better safe than sorry" guides our decisions to insure our homes and motivates us to get a second opinion when we face a risky medical procedure or consider buying high-waisted jeans. Yet, at other times, we also recognize the importance of risk-taking. "Fortune favors the brave." Or, in the case of hockey, Wayne Gretzky observed that "you miss 100 percent of the shots you don't take." Whether an unpredictable situation is seen as a risk or an opportunity depends as much on the facts on hand as on the story we decide to tell about them.

Writing in the 1982, anthropologist Mary Douglas and political scientist Aaron Wildavsky proposed a "cultural theory of risk." They argued that people's intuitive reactions to environmental or technological uncertainties stemmed from their beliefs about social organization.[68] Any person's knowledge of humanity's problems is unavoidably incomplete. We tend to paper over these otherwise gaping holes in our factual understanding with personal beliefs about the ideal organization of society. Libertarians' blasé attitudes about the uncertain risks of pollution have much to do with their commitment to avoiding constraints on individual freedom. Nuclear power can never be proven safe to some environmental leftists because safeguarding fissile material requires a high level of state secrecy and police power.

These cultural commitments also shape how we understand nature.[69] Some people are fatalists, seeing nature as inherently erratic (figure 3.3).[70] "Climate has always changed," they often say. "All we can do is adapt." Individualists are tempted to see nature as inherently stable and abundant, unable to be irrevocably harmed by human action. Or they might see nature as destined to be saved via as-yet-unrealized technological innovations. Hierarchists usually perceive risks to nature as manageable if kept within certain bounds, which justifies empowering experts to bureaucratically regulate it. Finally, there are the egalitarians, who view nature as inherently fragile and even precarious, lest people make the necessary lifestyle changes to live within planetary limits. Whether a person sees nature as something to ignore, manipulate for human benefit, shield from harm, or live with harmoniously depends on their worldview. How we ought to relate to nature doesn't emerge from "the facts" but rather finds

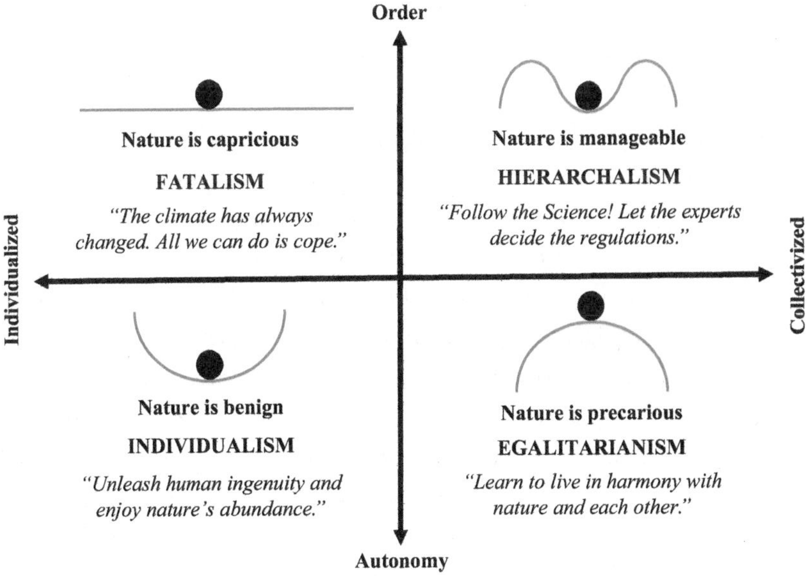

Figure 3.3

The four cultural theories of risk and associated myths about nature. The metaphor of a marble on a path represents each myth's assumptions about the resilience and variability of nature.

its roots in people's moral intuitions and personal commitment to certain ways of life.

Cases such as Guam and the brown tree snake seem to confirm the egalitarians' understanding of nature as fragile, a viewpoint that dominates the conservation movement. But other examples support a narrative of resilience. The cane-toad invasion of Australia looks like a catastrophe. But the species has become less poisonous over time. And the toad has not caused any species to go extinct, at least not across all of Australia. Many animals have adapted to their presence. Fish and other frogs have learned to avoid them, whereas black snakes have even evolved to tolerate toad toxin. Reflecting on decades of research, scientist Richard Shine argues that Australian ecosystems have not been "passive victims" to the cane toad. Even if it would have been best to have never introduced the toad in the first place, other species have proven to be surprisingly resilient. Shine concludes that "the impacts of cane toads are just one more disruption" in a dynamic system.[71]

Even invasion-species pessimist Daniel Simberloff acknowledges that losing half of Hawaii's native birds doesn't seem to have "caused natural or human-constructed ecosystems to malfunction"—because alien bird species have taken over dispersing the seeds of native trees.[72] But, ever the egalitarian, Simberloff also posits the possibility of "extinction debt." Various factors may be coalescing to drive a wide range of species to extinction in the near future. Perhaps ecosystems aren't resilient enough to adapt to invasions. It might simply be that not enough time has elapsed to see the damage. In contrast to Mascaro, Simberloff sees Hawaii as suffering from "invasional meltdown," as the effects of multiple alien species combine to open the door to still more nonindigenous species. Introducing nitrogen-fixing shrubs aids the spread of alien guava trees and earthworms, whereas feral pigs spread the seeds of non-native plants.[73]

Even if we embrace Simberloff's more dismal view of invaded ecosystems' future, there remains a silver lining. These failures beget learning. They remind us to proceed more prudently and stay vigilant. None of the species introduced as biological controls in the United States since 1965 have become invasive.[74] There are, no doubt, horror stories from the past. When Hawaii's Department of Agriculture introduced rosy wolfsnails in the 1950s to control the giant African snails imported decades earlier, they succeeded only in decimating native snail species.[75] But the individualist would remind us that, just as we have learned to make air travel and nuclear power plants far safer now than fifty years ago, biologists have figured out how to better cope with the complexities of species introduction.

In light of the frequent impracticality (and unintended consequences) of eradication, some advocate managed coexistence.[76] The role for conservationists would be making peace, not war. Can the interactions between natives and newcomers be mediated to create more stable ecological communities? For instance, some parts of eastern Australia are struggling with an invasive vine that smothers out tree growth. A local insect has adapted to eating the vine's seeds, but a cousin species from the north is even better at it. Hybridizing these two insect breeds would help keep the invasive vine at bay, creating a more even balance between native and newcoming species..

Could assisting the adaptation of already-existing Great Lakes species soften the potential impact of Asian carp? Would selective breeding, hybridization, or genetic engineering of South American beech trees make it unnecessary to cull beavers? Or might the reintroduction of the predators that humans may have helped exterminate thousands of years ago keep beavers from running rampant? Australian scientists, for instance, have already developed strategies to help impacted species overcome their cane-toad naivety.[77]

For "compassionate conservationists," ecological peacemaking aligns with their goal to "do no harm," even for invasive species. In this view, the decision to cull a species should only be pursued when all other options for coexistence have been exhausted. In his plea for hippo compassion, biologist Jacob Dembitzer argues for alternative strategies to mitigate the species' potential harms. For instance, simple fences made from chili peppers can deter hippopotamuses from invading farmland. In Dembitzer's view, the problem stems less from anything inherently wrong with the animal and more from our lack of effort to learn to coexist with large and often ornery wildlife.[78]

This view aligns with the perspective of some Indigenous groups—namely, the Anishnaabe in Michigan. The Anishnaabe understand alien species as migrating nonhuman "nations," a view that leads them to prefer less heavy-handed control measures and to ask themselves how they might best make use of the natural "gifts" that these newcomers may offer.[79]

Going even further could mean welcoming the invaders. In one particularly provocative passage of *Inheritors of the Earth*, Chris Thomas floats the possibility of introducing new flightless birds to New Zealand, such as rheas, emus, or Brazil's endangered Alagoas curassow, species that can deal with mammalian predators. Rather than continue to fight the losing battle to save mild-mannered birds such as the kakapo, why not bring in species that have a fighting chance? Thomas is quick to clarify that his idea requires careful study and experimentation (he isn't proposing "anything goes"). He nevertheless argues that it is more our own sense of guilt and "the presumption that the old species are better than the new" that leads us to fail to consider this option.[80]

Invasion biologist Mark Davis put it best when he argued that "non-native species force you to start making decisions about what kind of ecosystem you want and what your values are . . . They aren't a change from some iron-clad [environmental] truth, and it's not really scientific to act like they are. 'Unhealthy' or 'healthy' usually just means 'desirable or undesirable,' and we should be up front about that."[81]

Again, straightforward cases of unequivocally harmful invasive species no doubt exist. Chytrid's immense impact on global amphibian populations, for example, seems like an indefensible tragedy. Nor do I mean to imply that reasonable precautions, such as making cargo ships flush their ballast water tanks at sea rather than at port, aren't worthwhile.[82] Rather, my aim has been to show that the dominant "fragile nature" narrative regarding alien species obscures a more complex reality and limits our imaginations.[83]

Perhaps we might one day see the allergy to the hippopotamus as akin to the twentieth-century obsession with snuffing out wildfires. The belief that fires only brought devastation led to an attempt to curb them, causing even more destruction and less resilient forest ecosystems.[84] Nature doesn't always need saving from disturbances. Trees burn. Rivers flood. Species move. Hippopotamuses sometimes choke streams with feces. Disruptions play a role in ecosystems, even if the aesthetics can be offensive. Often, what is called "nature protection" is really about achieving a human ideal of a stable environment. Environmentalism in the Anthropocene must wrestle with the question of when to go to war on ecological disruptions and when to make peace with them.

SHOWDOWN

We still don't know who killed Pablo Escobar. He died in a shoot-out with police; that much is clear. But whose bullet ended his life? The answer to that question remains a mystery. Escobar's son insists that the smuggler died by his own hand, whereas the Colombian National Police eagerly take credit.[85]

Escobar's Nápoles estate is a theme park now. Had Escobar not succumbed in a hail of bullets, he undoubtedly would be pleased by what became of his former home. Advertised with a *Jurassic Park*–style font, the park's attractions include a full slate of mammals from the African savannah, a water park, and a hippopotamus named Vanesa, the proclaimed "Queen of Nápoles." Hippopotamuses have become a tourist attraction in their own right, bringing people to what would otherwise be an ignored rural Colombian river valley.

The government's patience for the animal has worn thin, just as it did for Escobar. In early 2022, the country's environmental minister added hippopotamuses to the country's list of invasive species. Plans are in the works for contraceptive and relocation programs. Officials worry about uncertain ecological harms should the population explode, along with the dangers that the several ton animals may pose to humans.[86] The decision turns the hippo into biological contraband, little different from the cocaine Escobar once trafficked through the country.

However, some locals are opposed. Álvaro Molina is one of very few Colombians who can say that they have been attacked by a hippopotamus (his home borders the river). But he is against eliminating them. Isabel Romero Jerez, a conservationist from the region, argues that "the hippopotamuses are already no longer African but Colombians."[87]

Jerez's defense of the cocaine hippos reminds me of the equally wickedly hard questions that often surround human migration. The question of when a species is no longer alien bears a resemblance to the question of when the descendants of Europeans in the Western Hemisphere would no longer be "settler colonialists" or at what point newly arrived immigrant groups become "American." Our ingrained tendency to project human cultural problems onto environmental disputes only makes the debate over invasive species thornier. Colombian hippopotamuses' impact on native fish could look like the Spanish conquest of Indigenous peoples, whereas more conservative citizens might think of them as akin to illegal immigrants outcompeting the native-born for jobs. When the science is already steeped in metaphor, these background narratives make interpreting the facts even more fraught.

I have a hard time not seeing the decision sterilize Colombia's hippopotamus population as premature, as an overly precautionary effort to ensure that the current ecological picture of the Magdalena River is frozen in time. It is fair to worry that the animal's disruptions will further degrade an aquatic ecosystem that is already stressed from human impacts. But should we forgo the possibility of learning that hippos might be benign, or even beneficial, because one group of conservationists believes that any potential for harm is intolerable?

Given that the South American continent has lost nearly all its megafauna, why not give hippopotamuses a chance and see how this natural experiment plays out? Some scientists believe that ancient humans armed with simple weapons had no trouble assisting the demise of similarly large herbivores. So, it seems unfounded to think that hippo populations could ever become so large as to challenge people's knack for extermination, especially when armed with modern rifles. Unlike Japanese knotweed, Australia's cane toads, and Patagonian beavers, hippos will never exist in such numbers that the decision to introduce them could not be undone. It is a trial whose errors can be corrected by run-of-the-mill poaching.

One could argue that the desire to see this natural experiment through is equally blind, the product of a romantic attachment to the idea of megafauna returning to their former stomping grounds. One set of scientists blamed hesitancy to cull the hippo herd as "social-ecological mismatch, . . . where the public perception or value placed on a charismatic exotic species constrains management options for removal or containing its spread in order to mitigate its negative effects."[88] However, given ambiguous evidence of hippos' harms and the complexities of the science on non-native species, blaming inaction on citizens' failure to "follow the science" is, frankly, dishonest. It hides precautionary environmental values behind a facade of factualness.[89]

Existing data are often insufficient to determine our decisions regarding what to do about a non-native species. This is why we invariably fall back on stories about the fragility (or resilience) of nature and humanity's place within it. Furthermore, we are faced with the realization that we are already and always helping to create novel ecosystems, not just through intentional introductions but also via climatic and habitat changes as well as our networks of trade and travel. Biologist Chris Thomas calls this the "New Pangaea," a melting pot of natives and newcomers in which only the more adaptable and resilient species will come to inherit the earth.[90]

We cannot escape difficult questions about what kinds of environmental interventions we should tolerate. As philosopher David Frank concludes, "The issue of what should count as *harmful* to ecosystems, the environment, or biodiversity . . . is an ethical issue that cannot be settled by ecologists alone."[91] Whether the hippos stay should not and probably cannot be answered by science, but only by thinking more incisively about how to better reconcile competing ideas about nature.

Maybe hippopotamuses really are ill-suited to the rivers of Colombia. Perhaps advocates of precaution are right. On the other hand, the more precautionary-minded may also be caught up in an inherited cultural attitude about newly introduced species. They survey an already artificial landscape, one divided into agricultural plots and long ago deprived of megafauna, and declare that several dozen hippopotamuses now threaten the ecosystem.

More and more scientists dispute the traditional views of invasion biology. One prominent ecologist told me that "by 2100 . . . everyone will wonder what the big deal was about 'invasive species.'" The debate could lead to a paradigm shift within biodiversity conservation. But such a paradigm shift may come too late for the Colombian hippopotamus.

Might hippopotamuses restore long-lost ecological function to South American waterways? Could currently naive species eventually adapt? If the Colombian government's plans are successful, the only certainty is that we will never get to find out.

4 ENGINEERING WILDERNESS

The boreal forest seems to sprawl out endlessly. Treetops of spruce, pine, and fir undulate in the freezing wind. It is a great green ocean whose waves can ripple and churn but never crash upon the shore. Moving north, or higher in elevation, forests eventually surrender to the tundra. Trees become gnarled and short, and moss and lichen now blanket the frigid ground. Short tufts of grass cling to life in soggy patches of soil.

Looking upon cold and isolated polar regions around the globe, it is easy to convince ourselves that they are exactly as early humans encountered them. However, some biologists contend that today's Arctic is an accident of history and a mistake to rectify.

Polar landscapes captivate us because we imagine them to be the last holdouts of pristine nature. The far north seems untouched, in part because human settlements are so small. Villages, towns, mines, and oil drilling sites are as stunted as the trees. They are hemmed in by the harsh conditions and dwarfed by the enormity of the terrain—although climatic changes are poised to open the northern latitudes to a "cold rush" as previously inaccessible areas thaw out.[1] According to the dominant story of the Arctic, it is only now that modern people imperil the world's polar regions. Their emissions will thaw the native permafrost, releasing untold tons of additional carbon into the atmosphere.[2]

But that story may be a myth. As we learn more about ancient human history, it appears to be inaccurate, if not misleading, to describe any of the world's ecosystems as natural. People's environmental impacts became global more than ten thousand years ago, long before *Homo sapiens* started mining fossil fuels out of the ground and burning them for energy.

When Russian geophysicist Sergey Zimov gazes upon the Arctic landscape, he sees the outcome of a mass extinction event. Neither the awe-inspiring boreal forests nor the greenhouse gas-releasing "carbon bombs" scattered within the permafrost should be there. It was only the demise of large mammals—mammoths, bison, woolly rhinoceroses, and cave lions, among others—that allowed trees to colonize the region and trapped incompletely decomposed plant matter in the ice.[3]

Zimov's solution is to "rewild" the Arctic. He cofounded Pleistocene Park in 1996 to do just that. The park is an ongoing scientific experiment that he comanages with his son, Nikita, and its goal is to gradually return megafauna to the remaining Siberian grasslands. The pair hope to eventually reintroduce lions and mammoths, if successfully resurrected. Zimov theorizes that by restoring the ecosystem of the Pleistocene, the geological era preceding the current Holocene era, not only will the Arctic become more productive and livelier, but we can also prevent the permafrost from thawing. Going back to a more ancient Arctic may be the key to securing humanity's future.[4]

Pleistocene Park is but one of hundreds of rewilding projects across the globe. These projects are motivated by the idea that we can do more than just restore a pre-1500 ecological baseline. By bringing back the biggest species, ecosystems can achieve new levels of vibrancy. Although it is certainly less gloomy than invasion biology, rewilding science cannot escape the same kind of quandaries. What exactly is the wilderness that rewilders hope to achieve? Where does the science of rewilding end and its environmental values begin? And where do people fit into a rewilded nature? Can they?

WHODUNIT?

When Pleistocene-era humans wandered the earth, they walked with giants.[5] Ancient Siberians lived among five-ton woolly mammoths and three-ton woolly rhinoceroses. In Australia, there were lizards longer than an SUV and marsupials that resembled giant capybaras. The Americas featured four-thousand-pound armadillos, ground sloths as big as elephants, and the short-faced bear, which was even bigger than today's polar bears. Where did they all go?

Two different theories compete with one another regarding what caused the demise of the majority of the planet's megafauna species: the overkill and the overchill hypotheses. The overchill hypothesis posits that climate change at the transition from the Pleistocene to the Holocene twelve thousand years ago was so rapid or disruptive that large species simply couldn't cope. However, one problem with the theory is that these megafauna species had survived previous Ice Ages and warming events. Why was the Holocene different?

A team led by University of Alaska scientist Daniel Mann believes that it wasn't climatic disruption that doomed the giants but rather the Holocene's unprecedented stability.[6] Large species have a very particular set of skills. They are better at turning food into energy, eat a broader diet, can tolerate fasting, and are more efficient at moving larger distances. The Ice Age's climatic instability threw landscapes into frequent disarray. Megaherbivores thrived because their diverse tastes and mobility were advantageous in a more chaotic and patchier environment. They could tolerate changes in food sources or migrate to greener pastures, whereas smaller, pickier, and more home-bound species would find that a sudden shift in the climate now left them without the plants they needed to survive.

Yet, in a more predictable Holocene climate, megafaunas' features became liabilities. They required a lot of calories and produced few offspring—and even then only after prolonged pregnancies. Smaller and more quickly reproducing species outcompeted the lumbering giants for food. And as megaherbivore numbers declined, the large cats and giant wolves also succumbed.

As plausible as the overchill theory sounds, the overkill hypothesis also has a lot going for it.[7] First, species extinctions seem to track human migration, beginning in Africa and preceding stepwise in Europe, Australia, and East Asia before affecting the Americas. And megafauna persisted where humans were absent. Woolly mammoths survived until just a few thousand years ago on some very remote Arctic islands. Second, extinctions disproportionately affected very large species, those that provided the greatest return on investment for human hunters. Some scientists reason that climate-driven extinctions should have affected a broader range of species.

The extinctions also varied in severity across the globe. In Africa, elephants, rhinos, hippos, giraffes, lions, and other large species persisted, whereas Australia and the Americas lost the vast majority of their larger animals. African megafauna might have survived because they evolved alongside early humans, whereas species elsewhere were naive to *Homo sapiens*. People probably also got better at hunting as they migrated.[8] By the time humans arrived in South America, they had been bagging megafauna for millennia.

The question regarding what wiped out Earth's megafauna eludes simple answers. We depend on an incomplete fossil record.[9] Did a species really die out twelve thousand years ago or is it that we simply haven't yet found more recent petrified remains? Finding fossils becomes increasingly unlikely as an organism becomes scarce, making the rapidity of extinction difficult to estimate. If people and megafauna coexisted for millennia before the latter finally succumbed, were humans really the decisive factor? These complications notwithstanding, many scientists who model megafaunal population decline claim that human migration better explains extinction than climatic changes.[10]

Knowing what exactly wiped out so many large animals at the end of the Pleistocene era might seem to be an esoteric scientific question. However, its answer changes the story behind rewilding. The overkill hypothesis makes people into an inherently destructive force, an intelligent species that started spoiling ecosystems before it had even figured out written language.[11] Even Indigenous peoples aren't off the hook, since nearly all of our ancestors would have played a part in slaying every last giant ground sloth or woolly rhinoceros. If megafauna were, on the other hand, simply unfit for a changing climate, then the case for rewilding starts to look less like environment restoration and more like experimental ecological engineering.

AN ARCTIC SERENGETI

The calm of the forest is pierced by the squeaks of moving tank treads, a familiar noise to anyone who has watched *Saving Private Ryan* or other World War II films. In this case, the squeaking comes from an old Soviet-era armored vehicle. The tank is a technological surrogate. It is a demonstration tool, a way to show visiting journalists how the landscape is altered by the mammoth's penchant for knocking down trees and compacting the snow.

Sergey Zimov is the tank's chain-smoking driver. He sports a beret, a bushy white beard, and an even thicker Russian accent. He believes that rewilding can make Siberia great again. Within Pleistocene Park's fences, Sergey and his son Nikita have introduced two kinds of bison, Yakutian horses, musk ox, Mongolian camels, and other animals. Gray wolves, Siberian tigers, and Canadian mountains lions could arrive in the future, but mammoths will have to wait for the necessary advancements in genetic engineering. The Zimovs hope to provide evidence that rewilding could not only fundamentally change the Arctic but also help mitigate climate change.[12]

The contemporary Arctic is relatively barren and desolate. In Siberia's extreme cold, new plant growth usually freezes before it can decay. Rather than adding nutrients to the soil, organic material is locked away in ice. This is part of the reason why the permafrost is full of carbon bombs, and it also partly explains the scarcity of life in these regions. In the tundra, mosses and lichen dominate. In the boreal forests, available nutrients are locked up in evergreen trees. Neither environment can feed large populations of herbivores.[13]

Buried deep in the permafrost, scientists have found traces of another Arctic, a far more vibrant version of today's Siberia. The frozen sediment is packed with evidence of plentiful grasses and the bones of diverse animals. Despite lower levels of atmospheric CO_2 and a colder, drier climate during the Pleistocene, the northern latitudes bustled with activity. Zimov estimates that every square kilometer could support a mammoth, five bison, eight horses, and fifteen reindeer.[14] How such a forbidding place could play host to so much life puzzled scientists, who referred to it as the "productivity paradox."[15] Putting that many herbivores in the current Arctic would surely lead to a barren, overgrazed landscape.

Zimov believes that megaherbivores' environmental disturbances resolve the apparent paradox. When their hooves trample the earth, they open holes in the moss for the earth to dry out and for grasses to colonize. Mammoths knock down trees not simply as a by-product of awkward lumbering but also to feast on their leaves, bark, and roots, as modern elephants also do. This keeps the forest from encroaching on the grassland that the large pachyderms helped establish.[16] Their grazing also speeds up decomposition. Herbivores are basically machines for munching up plants and defecating nutrients back into the soil. Pleistocene megafauna fundamentally altered the (sub)arctic ecosystem, engineering the environment simply by being themselves. The result was a unique grassland savannah, a much colder version of the African Serengeti, which Zimov refers to as "mammoth steppe."[17]

The maintenance of the grasslands ended with the megafaunas' extinction, leaving reindeer and moose as the habitat's new caretakers. Reindeer graze on grass, lichen, and mosses, whereas moose like to "browse," eating leaves, twigs, and bark. However, their activity isn't enough to sustain the steppe. What remains is an ecosystem that stores very little carbon in the soil, where it can support life. Instead, nutrients are locked away in permafrost or in forests, which are prone to burning down.

The consequences are not only more barren landscapes but also climatic disruptions. Grasses reflect sunlight better than moss, which helps keep permafrost frozen in the summer, whereas in the winter, the trampling of herbivores compacts the snow so that it cannot insulate the ground. This results in a deeper and longer-lasting freezing of the earth. Research by the Zimovs found that the ground inside the park was up to 10°C cooler than that outside of it.[18]

There is a certain appeal to the story of Pleistocene Park. A maverick scientist is cobbling together a radical ecological experiment that could help avert environmental catastrophe. It exemplifies the new conservation paradigm of rewilding, which can be seen as the antithesis of invasion biology. The story of rewilding is far more optimistic. It focuses less on the damage wrought by human meddling and more on the potential for a more vibrant and resilient nature. Even so, the science and practice of rewilding remain steeped in thorny complexities.

BACK TO THE FUTURE

Wilderness has a troubled history. Writer William Cronon argued, "We too easily imagine that what we behold is Nature when in fact we see the reflection of our unexamined longings and desires." The idea of wilderness has evolved alongside these longings. Wild nature's depiction as an area of desolation and danger eventually gave way to romantic portrayals of it as a place to "glimpse the face of God." Wildness is often thought as everything civilization is not, an Eden unblemished by human wickedness. Yet, these apparently uninhabited landscapes were often created by evicting their native peoples and erecting barriers to keep *Homo sapiens* out. Modern wilderness is both natural and artificial, a pristine environment that exists only through human intervention.[19]

Rewilding was born into this same contradiction. The term was coined by Dave Foreman, the cofounder of the radical environmental advocacy group Earth First! Foreman lamented the "wounds" that cattle grazing and wheat fields supposedly left on the environment and saw the potential to "rewild North America" in the bison herds of Theodore Roosevelt National Park.[20] Of course, doing so means repeating the process of park creation: fencing off large tracts of land and carefully reintroducing a mixture of species, especially predators. Foreman's proposal was formalized by scientists Michael Soulé and Reed Noss as the three Cs: cores, corridors, and carnivores, where land management policy builds pathways for species to migrate between large "cores" of protected area.[21]

Rewilding was further cemented in environmental scientists' imaginations by ecologist Josh Donlan. He advocated moving the baseline for environmental restoration in North America to roughly thirteen thousand years ago. Donlan argued that modern ecosystems still bore the mark of species that existed long ago. The North American pronghorn's speed, for example, is arguably the result of coevolving with a now-absent cheetah. The hope behind rewilding was that long-lost ecological processes would be restored by reintroducing species that had coevolved in the landscape. Donlan and his colleagues, moreover, saw the potential not just to enhance the continent's biodiversity but also to save threatened African species by establishing populations on the other side of the Atlantic. As with Foreman's proposal, Donlan's vision for a rewilded America required converting the "economically depressed parts" of the country's vast interior into giant parks.[22]

Rewilding is ultimately a tale of nature's debasement. It tells of an environment despoiled by humanity's original ecological sin: exterminating Pleistocene megafauna. Although only Foreman explicitly argues that organized agriculture sets humanity "apart from the natural world," both his proposal and Donlan's see much of America's vast agricultural lands as something to eventually repurpose for ecologically higher uses.[23]

There is something to these sentiments. Environments do seem less functional in the absence of megafauna. We have already seen Zimov's work on the Holocene Arctic's declining productivity. Megaherbivores' grazing and browsing keeps grasses trimmed and opens up dense thickets of trees and bushes. As a result, the large animals reduce the risk and severity of wildfires.[24]

Many of the environmental benefits of rewilding, however, could be summed up as "the power of poo." Megafauna help to cycle nutrients through an environment. Their loss has driven that cycle to a fraction of its historical value, on average a mere 6 percent of what it once was. Indeed, the decline of whale numbers has decreased ocean nutrient levels to the point that scientists are trying to synthesize whale poop substitutes. Whales' waste fertilizes the seas for phytoplankton, sequesters carbon, and supports fish populations. Ecosystems around the planet may be actually suffering from a deficiency in megaherbivores and their mega-dung piles.[25]

Although Pleistocene Park is certainly the most dramatic example of rewilding, not all rewilding scientists advocate something so dramatic as the return of mammoths to Siberia or importing rhinoceros to the American Southwest. Trophic rewilding sets its sights more modestly, looking to reintroduce megafauna and their ecological engineering without pining for a particular historical era. The trophic part emphasizes that the reintroduction of species leads to broad changes in food webs and feeding activities within the landscape. Passive rewilding involves letting abandoned agricultural fields lie fallow, letting nature decide how best to reclaim them. The latter is a more popular in Europe, where the environment has been intensively managed for far longer and wilderness areas are relatively scarce.[26]

Conceptions of rewilding and wilderness have also evolved. Perhaps it is more accurate to say that they returned to their roots. Rod Nash argues that "wild" was originally a contraction of "willed," that "wilderness" meant "self-willed" lands.[27] That is, such places are not steered by people's desires but rather driven by the nonhuman species that take residence within them. This shifts the aim of rewilding to one of "reorganisation rather than restoration," introducing species with the hope of achieving a self-sustaining and productive ecosystem. Rewilded lands should no longer need intensive human management.[28] Although early advocates looked to the Pleistocene era for inspiration, rewilders now recognize that they are building novel ecosystems.[29]

The radicality of rewilding has attracted detractors. The biggest concern is that reintroductions will bring unintended ecological consequences. Some scientists go so far as to call rewilding a "Pandora's box."[30] On the other hand, traditional conservation approaches have led to their own unintended consequences. Policies encouraging people to only bag full-grown rams or snag larger fish have had evolutionary impacts. Now, the rams grow smaller horns, and the fish never seem to fully mature.[31] Although the critics have a point, rewilders contend that the risks are more manageable in comparison to other radical environmental proposals, such as spraying aerosols into the atmosphere to control the climate. They call for continuing to develop the new ecological paradigm's scientific basis.[32] But as we will see, even the most iconic examples of rewilding have not exactly reduced the idea's scientific uncertainties.

IF I COULD TURN BACK TIME

A baby bird flaps its wings in distress. The fledgling is being stalked, but not by a hungry cat. Lumbering toward it is a several-hundred-pound Aldabra tortoise. The reptile slowly stretches its wrinkly neck and opens his mouth. The scene oddly reminds me of how an old, turtle-neck-wearing professor of mine looked when eating a burrito in class. Then, in a distinctly nonprofessorial manner, the tortoise's jaws snap shut, crushing the bird's skull.[33]

Nature often surprises us. Seemingly nonthreatening animals can turn out to be predators. And restoration efforts can fail to return ecosystems to their past state.

On several small islands belonging to the Republic of Mauritius, giant tortoises were once the biggest kids on the block. At first, scientists thought they were a result of "island gigantism," a case where the lack of predators allows isolated species to grow to huge proportions. Classic examples include the large flightless birds of New Zealand, such as the now-extinct twelve-foot moa. However, giant tortoises' former stomping grounds included mainland Africa, so the species' evolutionary path remains a mystery.[34] In any case, Mauritius's behemoths were driven to extinction by the 1840s. Even so, the environment continues to bear their mark. Palm and ebony tree species there grow large fruit with hard seeds. They coevolved alongside megafrugivore tortoises, who could swallow the seeds and disperse them (slowly) across the landscape. Because of their absence on Île aux Aigrettes, ebony forests have failed to bounce back after the end of logging.[35]

In the 2000s, conservationists released Aldabra tortoises from small enclosures and into the wild. Although this species it not native to Mauritius, scientists believed that it could replicate the ecological function of the now-extinct giant tortoises, acting as a "taxon substitute." Scientists hoped not only that the newcomers would help native trees bounce back but also that their grazing would suppress non-native plants that were, by then, widespread on the islands.[36] Harm to baby birds doesn't seem to have been on researchers' radar, but one presumes that only the most fleet-footed of tortoises manage to snag a feathered snack.

The results of this rewilding have been mixed. The tortoises have been eating tree fruits and spreading the seeds, but the reptiles haven't had a big impact on controlling alien species or bringing back a native mixture of plants. In fact, there is some evidence that Aldabra tortoises consume and help disperse exotic species.[37] Although conservationists have returned an iconic part of the Mauritian landscape, the ecosystem itself will remain a hybrid, a modern creation.

Contrasting the Aldabra tortoise's immensity, tule elk are North America's smallest wapiti subspecies. They are also the most controversial. These ungulates were once a staple of the Miwok people's diet but suffered a similar fate as the buffalo that used to fill the Great Plains. The Miwok now live on rancherias and are scattered throughout northern California. The tule elk are similarly dispersed. A mere six thousand elk roam the state's parklands. Finding a place for the species in the modern Californian landscape has been contentious.[38]

Restoration of the tule elk to Point Reyes National Seashore hasn't exactly returned the landscape to the past. As with the Aldabra Tortoise, the ungulates happily munch on imported plants as much as native ones. That said, they have steadily reduced the coverage of non-native velvet grass on the landscape. And they have also reduced shrub growth, opening up grassland habitats. Yet, they can't do much about the cows that graze nearby.[39]

Studies find that cows and the elk mostly keep to themselves. It is the coexistence between ranchers and conservationists that is uneasy.[40] Point Reyes only became a national seashore in the 1960s. Getting the land into the hands of the National Park Service required congressional involvement. The result was a compromise between the park service, local ranchers, and dairy farmers, allowing the latter groups to obtain grazing permits and leases in designated pastoral zones.[41]

When those leases were up for renewal in 2012, environmental groups sued to try to stop them. The legal battle continues today. The park service wants to cap the northern elk herd's size and reform grazing practices. The Center for Biological Diversity, on the other hand, demands that agriculture completely cease within Point Reyes and that fences be removed in order to ensure "maximum protection, restoration, and preservation of the natural environment." The ongoing conflict is acrimonious, to say the least, with environmentalists calling the park service's plan "insane" and "greenlighting the slaughter of native wildlife."[42]

The herd at the center of the conflict was itself an environmental surprise. Tule elk were settled in a wilderness area on the opposite side of the estuary from the area's farms, but a few adventurous animals swam across it and made an amphibious assault on Drakes Beach.[43] The park service's proposal entails culling a handful of just these elk, not the herds in the wilderness areas. Given this context, the Center for Biological Diversity's claims about "slaughter" are definitely exaggerated. That said, the battle isn't really about a few culled elk but rather about what rewilding represents to many of its advocates: restoring an ecological "masterpiece" by only preserving what "has belonged there."[44] However, not everyone agrees what the ecosystem should look like. When asked to compare the rewilded areas with the rest of Reyes Point, Hall Cushman, a scientist who has studied the tule elk since their reintroduction, responded, "Is one better than the other? That's a tough call."[45]

Rewilding in both Mauritius and Reyes Point illustrate the challenges of trying to wind the clock backward. A scientist tracking the elk put it clearly: "You can't take a piece out and expect it to be the same way it was when you put it back in."[46] Not only have new plant, animal, and insect species moved in but so have humans, along with their pets and livestock. Even if we can't transport a habitat back in time, rewilding still promises another kind of a restoration: returning order to degraded ecosystems.

THE BALANCE OF NATURE

"The rivers changed in response to the wolves," intones journalist George Monbiot in a now viral online video. Monbiot breathlessly describes how bringing the wolves back to Yellowstone demonstrates the miraculous impacts of rewilding. The predators have "managed" the elk, allowing trees to regrow in the valleys, stabilizing the riverbanks. Habitat proliferates for songbirds. Beavers build more dams. Scavengers have more food. Ecological balance has returned to the park. The video cuts to a wolf howling under a rainbow while sentimental synthesizer music plays in the background.[47] The aesthetics are melodramatic and the story oversimplified.

In scientific terms, wolves are hypothesized to have instigated a "trophic cascade" within Yellowstone. That is, their impacts are not limited to their prey but are believed to propagate across multiple levels of ecological organization and down the food chain.[48] In the case of Yellowstone, wolves predate the elk, which spares grasses and trees and provides carcasses for scavengers to eat. This, in turn, has effects on fish, beaver, and other wildlife. A park thought to be dangerously off kilter, with elk overgrazing its northern range, now seems to have leveled up to new heights of vibrant biodiversity.[49]

Proving that story is no easy task. Elk populations have dropped since wolves were reintroduced in 1995. Even so, that decline also parallels increases in bears and cougars, more elk hunters, and recurrent droughts. Studies find that aspen trees and berry bushes are rebounding. Some researchers credit the changes to the creation of a "landscape of fear," a situation where the threat of wolf predation leads elk to be too nervous to eat the baby trees sprouting in exposed areas.

Whether the wolf is actually responsible for the new Yellowstone is heavily disputed. Like other cases of environmental science, the choice of metric can be decisive. It looks as if elk are being kept away from new aspen growth if one measures the *tallest* tree in the stand near wolf killings but not if one takes the *average* height.[50] Still, it is unlikely that the wolves do nothing. In the far simpler island ecosystem of Isle Royale, Michigan, the peaks and troughs of wolf and moose populations are exactly opposite one another.[51] For larger and more complex ecosystems, however, authoritative data are elusive.

The viral story of Yellowstone leaves out what the wolves alone cannot change. Willow trees needed a similar respite from elk browsing if they were to reestablish themselves. Another necessary condition for their regrowth is that beavers flood the nearby environment through their dams, which brings the necessary sediment for the willows to take root. Unfortunately, this is a catch-22. Beavers need access to tall, straight willows in order to move back in and build the dams that would, in turn, help willows recolonize the park.[52]

The narrative of a triumphantly rewilded Yellowstone persists despite these complications. The story builds on the metaphors behind many people's understanding of the environment. Some think of nature as a divine thing, a godly order. Others see it in terms of a mechanistic system, a kind of biological clockwork. In either case, there is an assumed balance to nature that is disrupted (or alternatively restored) by human action. Nature, under the right conditions, is supposed to achieve a state of constancy, a "natural ecological climax." Wolves seen through this metaphor become a keystone species, a vital link in restoring balance and stability to Yellowstone. Accumulating evidence over the twentieth century, however, has only illuminated the environment's randomness. The natural world is ever-changing. It seems infinitely capable of bringing about and weathering new disruptions: fire, drought, inundation, disease, and so on. Prey species rise and crash in response to all of these factors, not just management by apex predators. Wherever ecologists have looked for an ordered stasis, they have found a surprisingly complex dynamism.[53]

In an era when we only hear about impending environmental catastrophe, positive environmental stories can seem appealing. Many areas of conservation, such as invasion biology, have traditionally portrayed nature as defenseless and innocent. Environmentalists' role is thus limited to resisting the onslaught of human-driven environmental degradation. In contrast, rewilding shines a ray of hope. It offers the potential for people to help unlock and recover new ecological dynamics. Conservationist Paul Jepson calls this story "Recoverable Earth," and he believes it can serve as an antidote to depressing tales of catastrophe and promote an environmental "awakening."[54]

That hopeful story can easily turn sour. We risk overselling rewilding's benefits whenever we get caught up in metaphors of nature as a carefully balanced order. "Recoverable Earth" poses similar risks as "invasion." It is a vision that can get recast into a seemingly undeniable truth. Once transformed, the scientific and social complexities of recovering nature fall from view. They are replaced by a more single-minded drive to assure ecological progress. George Monbiot, for example, now calls farming "the number one threat to the planet." He argues that rewilding agricultural fields and shifting people to lab-grown foods are imperative to save nature.[55] A redemptive vision of a bright ecological future inevitably casts a shadow on all that fails to rise to its lofty heights.

The challenges for rewilding are not only scientific. We should remember that wolf reintroduction was a political nuclear blast, a vicious battle whose aftershocks continue to reverberate throughout the American West.[56] Conservation's opponents saw wolf restoration as a case where urban environmentalists imposed their ideal vision of wilderness on everyone else. The pathway to a wilder world unavoidably winds through humans' diverse demands on and desires for nature.

MAN-MADE WILDERNESS

The grassland is littered with corpses. One captive slowly walks by, their eyes sullen, and with ribs protruding through sagging skin. Others crowd the fences, hoping to find a way out. Protestors manage to force a gate open. They want to bring food to the starving inhabitants, but local police drive them back. In the protestors' view, the facility is the moral equivalent of Auschwitz, the infamous Nazi concentration camp. However, to managers of the Oostvaardersplassen (OVP), the mass starvation of horses, deer, and cattle is just nature taking its course.

What is now the OVP once sat under the sea. Over the twentieth century, the Dutch have erected massive engineering works to reclaim lands lost to waves and floods. One of the most awe-inspiring is a twenty-mile dike that holds the North Sea at bay. Decades of draining have resulted in the creation of whole new provinces. The OVP was originally slated for industrial development, once it was sufficiently dried out. Instead, the area developed into a surprisingly rich wetland ecosystem.[57]

When Dutch biologist Frans Vera visited in 1978, what he found was eye-opening. Migratory geese seemed to be maintaining the wetland for themselves. The shallow marshes would get choked out with reeds, only to be opened up again by the geese's grazing. However, trees were beginning to encroach the OVP. Vera saw an opportunity to conserve the newly developed landscape. Through his job at the Dutch forestry service, Vera imported the closest equivalents to Europe's prehistoric grazers. Konik horses were introduced in place of the diminutive wild tarpan (figure 4.1).[58,59] And aurochs were substituted by Heck cattle, a product of an early twentieth-century selective breeding experiment supported by infamous Nazi Hermann Göring.[60] Red deer were also introduced from Scotland. As the geese

Figure 4.1
Konik horses graze near wetlands in Oostvaardersplassen.

had done for the reeds and marshes, reasoned Vera, so too would the megaherbivores do for the trees and grasslands.

Asked what an undisturbed environment looks like, many people would conjure up images of a primeval forest, something dense, dark, and dotted with dead logs. Vera argues that this view of "pristine" nature is misguided, a product of generational amnesia blinding us to what the landscape once was.[61]

Vera contends that a landscape's vegetation is always in a cyclical process of transformation. Mature closed-canopy forests eventually collapse, forming glades. New trees and grasses sprout in the opening, attracting bison and other herbivores. Their grazing and browsing expands the glade, which eventually forms a grassland. But even this state isn't going to last. Tree seeds find refuge under thorny bushes. Young forests grow and mature as the cycle starts anew. The result is a landscape that isn't all forest or all grassland but rather a dynamic and diverse mosaic of habitats. It is a patchwork of lands in various stages of growth and collapse.[62]

Vera's view of the (prehuman) European landscape is disputed by some paleobiologists, who point to the lack of grass pollen in the sediment cores that they drill out of marshy depressions. Others counter that grass pollen was only scarce because it was eaten by herbivores. A study on beetle fossils argues that species feeding on herbivore dung were far more prevalent in the Pleistocene, suggesting that grasslands used to be more widespread. Although the scientific debate rages on, Vera now deemphasizes pining for a long-lost Europe, instead arguing for the creation of more vibrant ecological relationships.[63]

The OVP experiment was an opportunity to test Vera's ecological theory. Whether it has been a success or a failure depends on whom one asks. Vera certainly thinks it is working. Thousands of birds and herbivores now roam the area, maintaining the grassland, and thorny bushes have sprouted in the infrequently grazed parts. Nonetheless, populations of cattle, horses, and deer dramatically boom and crash. During some of the worst years, thousands of grazers have either starved or been shot by managers, who picked out the weakest individuals with the "eye of a wolf."[64]

The hands-off approach applied at the OVP runs contrary to the usual aesthetics of pristine nature. Yet, the bounty offered by hundreds of deer, horse, and cattle carcasses isn't really different than that of a field of lush grasses. Foxes, ravens, and a pair of rare white-tailed eagles happily gorge themselves on the abundant carrion. Even though such a scene is no less emblematic of wild nature than the Yosemite Valley, almost no one would buy a romantic landscape painting of it.

Environmental changes at the OVP also seem alarming when viewed through traditional conservation measures. Some areas look overgrazed, like a wasteland, but ecologist Han Olff insists that people miss the bigger picture: "They don't see the landscape variability." When reedbeds are heavily grazed by geese, reed warblers might shift to the marshes. Still other bird species then move in, mostly those that prefer tightly grazed areas. Spoonbills were thought to have been wiped out by foxes, but it ends up that they had taken up residence in the wider area. Countless trees have died under the pressure, now littering the landscape, but more resilient species are coming to replace them.[65] A rewilded nature, in this view, simply features a lot more death and disruption than we are used to. We are mistaken when we equate the cycle of disturbance and regeneration with ecological degradation.

Frans Vera contends that our perception of death is also the product of generational amnesia. Our standard for animal welfare is based on what we see in farms, zoos, or safari documentaries featuring lions pouncing on passing gazelles. In contrast, only a minority of wild herbivore deaths are due to predation. Three-quarters of African wildebeest, for instance, succumb to malnutrition, not to teeth and claws.[66]

On the other hand, the OVP isn't the Serengeti. It is a twenty-two-square-mile reclaimed polder turned into a nature preserve, located fifteen feet below sea level and fenced in by farms, dikes, and a rail line. The water level is carefully controlled to maintain wetland conditions. Organizers offer safari-like tours for $45, where visitors can snap photos of untamed animals and crop out the high-voltage power lines when they get home.[67]

Vera maintains that others misunderstand wild nature, but he perhaps misunderstood his own experiment. One set of biologists argues that the OVP failed to satisfy the three Cs of rewilding mentioned earlier in this chapter.[68] It was too small to be a core protected area. Attempts to build wildlife corridors (or pathways) to other nature preserves failed, which meant that growing herds couldn't migrate to greener pastures. And carnivores were absent, forcing hunters to serve as surrogate predators.

The OVP arguably also ran afoul of a fourth "C" of rewilding: compassion.[69] To turn a blind eye to the suffering of animals intentionally brought in for an experiment in nature restoration seems callous. It is a moral failure that cannot be absolved by pointing to rewilding's commitment to animal autonomy. Even so, if some kind of wilderness is the goal, humans can't and shouldn't try to eliminate all animal suffering. Philosophical proposals to treat other species as morally equivalent to humans typically end in impractical and seemingly ridiculous conclusions. Martha Nussbaum, for instance, suggests that we deter wolves and lions from hunting and killing prey, supplying them instead with "humanely killed meat or fish . . . or lab-grown meat."[70] It is a proposal that runs to the opposite end of the spectrum, asking humans to lord over nature as moral schoolmasters. In any case, as conservation efforts continue to blur the line between wilderness and zoos, we will be confronted with difficult questions regarding humanity's responsibilities to other species.

Vera continues to insist that herbivore populations would have eventually stabilized. Insisting is all he really can do, having been largely pushed out of decision-making about the OVP. The Dutch will have to learn to navigate the difficult terrain of man-made wilderness without him.[71] As rewilding transforms from a radical conservation proposal into real-life restoration practice, cracks form around what otherwise seemed to be straightforward ideas about the proper place of people in the environment.

LEAVING IT UP TO NATURE

Gas masks litter the floor of the rundown school house. Most of the area's buildings now lack windows. Each one looks to be in a different stage of surrender to nature's steady forward march. It could be a scene dreamed up in Hollywood, the setting for a postapocalyptic action flick. The area's apparent bleakness is contradicted by a resurgent nature. Bears raid abandoned farmhouses, moose and elk are seen browsing a lush landscape, and the feral descendants of village housecats catch field mice to feed their young.[72]

Human negligence made roughly one thousand square miles around Pripyat, Ukraine, into Europe's third largest nature reserve (figure 4.2).[73] After

Figure 4.2
Nature is slowly reclaiming the formerly human-settled parts of the Chernobyl Exclusion Zone.

a reactor at the Chernobyl nuclear power plant caught fire and exploded, Soviet officials established an exclusion zone to keep people away from the highly irradiated landscape.

Things initially looked grim for the surrounding environment. A thousand acres of pine expired under a blanket of radioactive particles. The area is now called the Red Forest, referencing not communism but rather the rusty color of the dying trees. Many wondered if and how the environment would recover. Studies in the high radiation areas by an American-Swedish team of scientists found swallows with malformed tails and beaks, infertile birds and voles, and overall lower levels of species abundance.[74] Nevertheless, other researchers disputed these dismal findings. Farmland species have declined, but others, such as the greater spotted eagle, seem to be on the rise. Wetlands have expanded, partly in response to beaver activity. Amphibian populations remain high, with some species apparently developing darker skin as a defense against the radiation. Meanwhile, researchers in the Belarussian half of the zone contended that wildlife populations were no lower in irradiated areas than elsewhere. In response, a journalist for *Science Magazine* concluded, "Humans are worse than radiation for Chernobyl animals."[75]

The controversy is magnified by the politics of nuclear technology. The effects of chronic low-level radiation exposure on humans are uncertain and highly disputed.[76] Chernobyl's wildlife is a literal canary in the coal mine. The region's surprising natural vibrancy could be used to justify allowing people back in. Indeed, that rationale seemed to be behind the Ukrainian government's decision to open the region up to tourism in 2011.[77]

The Chernobyl Exclusion Zone is a natural experiment in "passive rewilding." In contrast to the cases discussed so far, passive rewilding involves ceasing to intensively intervene in a landscape. Much of today's passively rewilded lands were once agricultural fields. Marginal farms in Europe and North America are increasingly abandoned in the face of global competition. According to Portuguese scientist Henrique Pereira, passive rewilding entails three steps. First, wildlife is allowed to return by restricting hunting or relocations. Landscapes are then permitted to reconnect, most likely by further land abandonment. And third, people learn to tolerate the return of disturbances, such as fire, pests, and floods. As Pereira puts it, "If you love it, set it free."[78]

The hypothesized ecological process underlying passive rewilding is similar to Vera's theory. By allowing the return of natural disturbances, the landscape becomes patchier and more diverse, more "heterogeneous" as scientists put it. Browsing by bison, the digging of boars, small wildfires, and other small disruptions increase local biodiversity by creating a mosaic of ecological niches.[79] The return of these environmental processes offers additional benefits to humans. Carbon is sequestrated, soil erosion is reduced, and water quality is improved as tree numbers increase and farm runoff declines.[80]

Although land abandonment is obviously lower stakes than reintroducing big carnivores or unruly Heck cattle, it remains controversial in rural areas. Passive rewilding would no doubt strike nature lovers from the city as an unmitigated gain because only rural residents experience the negatives. Rural lifeways disappear, along with knowledge about traditional plants. People living in these regions feel left behind, as if it were not only the fields that were being abandoned but also themselves and their heritage. A study of Mediterranean farmers found them to be hesitant to rewild, only willing to do so if paid a thousand or more euros per year for every soccer field–sized patch of land that they abandoned. They name a high price not because of the income lost but rather because of the emotional costs of moving away from their traditional way of life. Rewilding also runs contrary to their aesthetics. Farmers take considerable pride in having "tidy" fields.[81]

Passive rewilding's apparent simplicity can obscure challenging complications. Land abandonment can simplify local ecosystems, and even reduce biodiversity, as non-forest species are pushed out by advancing trees. Such has been the case in Japan. Numerous species thrived in the artificial wetlands of farmers' rice paddies and in the grasslands created by clearing woods. They now struggle in the absence of farmland. Looking upon America's expansive fields of corn, we can forget that agricultural landscapes need not be biodiversity deserts. Traditional farms were often themselves a diverse mosaic of fields, forests, and pasture. They offered a mixture of habitats and transition zones that supported a wide variety of wild species. Although passive rewilding aims to eventually be hands-off, it is a major challenge to set the landscape on the path toward a similar kind of mosaic pattern.[82]

The Chernobyl Exclusion Zone has been unaffected by many of these complications. It wasn't conservation but rather catastrophe that drove people from their homes. And that is probably why it is such an appealing case for some environmentalists. The nuclear disaster precipitated a kind of emergency politics that some people wish could be applied to biodiversity. That said, management of the zone hasn't exactly been hands-off. Przewalski's horses were introduced in the 1998. Nevertheless, some see the Chernobyl Exclusion Zone as a model for how humans ought to relate to nature: Leave it alone.

THE CALL OF THE (RE)WILD

For all the media interest and documentary filmmaking, Pleistocene Park remains a tiny research reserve. Progress has been halting. Moving a handful of horses, bison, reindeer, and other animals to the remote compound has cost hundreds of thousands of dollars, and scientific collaborations have been thrown in disarray by the Russian invasion of Ukraine. Landscape changes have been slow. Animals have shunned consuming woody vegetation, and there may not be enough grass seed in the soil to sprout in the disturbed ground, leading managers to consider seeding and fertilizing grasslands to kick-start the process.[83]

Initial scientific results seem promising, although other scientists have criticized several methodological flaws that limit the research's authoritativeness. One study found higher carbon turnover within the park. In other words, carbon was being absorbed by the plants and soils more quickly and then rereleased into the environment. Moreover, less methane escaped the permafrost because the drier topsoil kept it from thawing so deeply. These results seem like a big step toward confirming Zimov's hypothesis: Grazing increases the productivity of the land and stabilizes the permafrost.[84] Because the Zimovs lack the resources to replicate ancient Siberia on a large scale, however, producing unambiguous conclusions will remain a challenge. Pleistocene Park may be most powerful as a symbol. The experiment represents the opportunity to recover a more resilient and vibrant earth.

The same is true of rewilding more broadly. Every component, from the overkill of Pleistocene giants to the restoration of trophic cascades and the reestablishment of lost ecological processes, is scientifically plausible. There is supportive, albeit disputed, evidence. However, conservation is as political as other areas of life. Every dollar going to rewilding is potentially one not going to invasive species control or to shore up rural economies. Rewilding's risks aren't only ecological but also social. It doesn't help the situation that the story behind rewilding, the narrative of creating reserves of self-willed nature on an Anthropocene planet, doesn't seem to have much of a place for people, at least not yet. After a suitable number of places are rewilded, where does the human species, the one turning the ecological dials and gauges back toward wilderness, fit in?

As I have tried to show over these three chapters, biodiversity science can't escape storytelling. Scientists in conservation biology are not necessarily more prone to it than elsewhere. And really, they have no other option. The subject matter is so awesomely complex that it eludes people's best efforts to develop decisive facts and authoritative interpretations. A person would need to talk to a hundred scientists to begin to grasp all of the intricate relationships between soil microbes and geochemistry, plant growth, animal behavior, and atmospheric conditions that underlie Sergey Zimov's proposal for a rewilded arctic. Even then, they would hear conflicting facts, incommensurable theories, and divergent interpretations.

Likewise, for non-native species, no one can say how an ecosystem colonized by a wide array of newcomers will look in a hundred years' time and why it should matter to planet Earth. Nor is it straightforward to deduce what the consequences of a decline of, say, 10 or 20 percent in the global catalog of species would actually be, much less decide the extent to which people should care.

Simply put: We need stories about the environment. Otherwise, it is too much for anyone to understand. We swim in a sea of competing and unavoidably biased visions of nature, only rarely and briefly feeling the bedrock beneath our feet before the crashing waves pull us back into the churn. In light of the seeming urgency of global ecological problems, it is tempting to dogmatically insist that *our* environmental story is firmly and uniquely anchored to objective reality. It is the others who are adrift.

Decisions about nature are also politically fraught. The price of environmental negligence can be the loss of species that people find culturally valuable or whose uniqueness leaves us in awe. But succeeding in conservation, as we have seen for rewilding, often places burdens on the people who make their living from the environment. Contention only partly or indirectly stems from differing ideas about the fragility of nature, or how pristine we think it should or could be. Mainstream environmentalism is controversial because it is often incompatible with what other people want from the environment. A landscape full of mammoths and tigers doesn't leave much room for wheat fields and dairy cows. As writer William Cronon pointed out decades ago, our stories involving nature are really about us, about how we think humans ought to live. They answer the question: Who should get what, when, where, and how from the environment?

For all their differences, both invasion biology and rewilding emerged equally steeped in the kind of wilderness thinking that Cronon criticizes. Regardless of whether it is longing for the species mix that existed prior to human settlement or looking to a thriving and self-sustaining ecological future, conservation struggles to shake its view of humanity as inherently polluting the environment. "The tree in the garden [is] artificial—completely fallen and unnatural—and the tree in the wilderness [is] natural—completely pristine and wild."[85] Ecological harmony appears to exist only where we do not, leaving us duty bound to practice environmental self-abnegation. We come to believe that it is our responsibility to exhaustively chronicle the injuries we impart on nonhuman species and to somehow restore the land to what it would be without us.

Biodiversity scientists are increasingly self-reflective when it comes to wrestling with this paradox. In science journalist Emma Marris's book, *Rambunctious Garden*, we are introduced to researchers reconsidering the traditional animosity toward non-native species, becoming open to novel ecosystems, and thinking about how to create wilder spaces without a commitment to pristine nature or historical baselines. Sergey Zimov, for one, seems to relish the role of being a rambunctious gardener of wilderness: "Playing God doesn't bother me in the least. We are already doing it. Why not do it better?"[86]

I will return later to the question of what it might mean to pursue a different idea of wilderness. Must storytelling about nature place humans in the role of either ecological wrecking balls or enlightened environmental restorers? In the next chapters, I turn away from the scientific and philosophical difficulties of biodiversity conservation in order to look more closely at its politics. Nature and wilderness are far from the only values that we must contend with. People care just as much about protecting their livelihoods and traditional cultural practices, feeling included in collective decisions, and everything else that makes life worth living. Realizing a wilder Anthropocene will be impossible without learning how to more productively navigate conservation conflicts.

5 DIVIDED

Save the Cowboy, Stop American Prairie Reserve.

Signs with this slogan now dot the grasslands of Montana. It is the most evident evidence of the resentments bubbling beneath the surface of an otherwise bucolic rural landscape. At the center of the controversy is an East Coast biologist, a former Silicon Valley entrepreneur, and their half-a-billion-dollar plan to bring buffalo back to the American Great Plains.

That entrepreneur, Sean Gerrity, was originally raised in Great Falls, Montana. Upon retiring in 2000 to Bozeman, after a long career in California, he and biologist Curt Freese founded the American Prairie Reserve (APR). They weren't the first to propose de-cattling and restoring bison to the Western United States. Two urban planners in 1987 argued that we ought to return the plains "to its original pre-white state." Gerrity and Freese, however, were the first to have the money to start doing it. They have raised more than $100 million from wealth philanthropists, wooing them with visions of a restored western ecosystem. APR has used those funds to buy up a hundred thousand acres of private land, a fifth of the way to their long-term goal. By stitching together private holdings with wildlife refuges and grazing leases, Freese and Gerrity hope to secure five thousand square miles for the buffalo to roam.[1]

One person's dream of a restored wilderness is another's nightmare, a dreadful vision of hostile annexation. Many ranchers see malicious intent beneath APR's glossy photos of grazing buffalo and behind their attempts make peace with locals. "I think the prairie reserve's endgame is to depopulate this area," says one area resident. In their view, restoring the native landscape will bring about the regional extinction of not just cattle but also ranchers themselves. Residents have become so mistrustful of APR that they have even banded together to buy up any farms targeted by the organization. They have put these lands under "negative bison easement." By forbidding buffalo from these parcels for a period of twenty years, they hope to assure a future for ranching.[2] This reaction might seem disproportionate, a fearful knee-jerk response to changing times and a few dozen ranches changing hands. However, digging into the story behind conservation conflicts reveals the complex and understandable reasons behind rural resentment.

Land in North Central Montana, like most elsewhere, is a political problem. Competing interests collide in even relatively mundane decisions over otherwise desolate plots of dirt. Maps of the Western United States resemble a quilt, a colorful patchwork of ownership and use designations. Cut a straight line for a hundred miles in any direction and a person crosses not only private properties but also an alphabet soup of governmental holdings. Authority can rest with the Bureau of Land Management, US Fish and Wildlife Service (FWS), the Park Service, or Native American reservations, among other organizations. To make matters worse, everyone wants something different from the land. Some see the West as a place to extract oil and gas, to log timber, or to raise cattle and a family. Others view it as a place of recreation, for rafting, snowboarding, and birding. Western policy has long been occupied with negotiating the coexistence of these varied uses, but only some goals for the land are compatible with restoring ecosystems to their pre-European appearances.[3]

APR is a prelude to future biodiversity conflicts. Should nations around the globe follow through with their Montreal–Kunming commitments to protect 30 percent of the earth, I predict that political hostilities will surpass what we've experienced for climate change. That is, it will unless we are willing to learn from our past experiences and to strive to better understand the roots of conservation disagreements. Why do reintroductions provoke such ire? How come victories by environmentalists only seem to provoke more political polarization? Who suffers which benefits and which harms when it comes to conservation, and who gets to decide?

"In all honesty, I don't think this is about grizzly bears. This is about power." That is how one landowner responded to a plan to reintroduce bears to the Bitterroot Mountains of Montana and Idaho.[4] And it is a fitting response. Beneath objective-sounding language about ecological functioning lies a discomforting reality. Conflicting visions of nature empower different groups of people to speak on its behalf—and to enjoy its economic and cultural benefits.

Division in the American West runs deep. Environmental philosophies had already polarized by the turn of the twentieth century. "Conservationism" originally meant slowing down the pace of extracting natural resources to a sustainable level. The goal was to promote the "wise use" of trees, water, land, wildlife, and other environmental gifts so that future generations could also enjoy their benefits. President Theodore Roosevelt and his chief forester, Gifford Pinchot, cemented its principles into American policy. Modern environmentalism, in comparison, draws upon what used to be called "preservationism." Early proponents, such as John Muir and Aldo Leopold, championed an ethic wherein the land was more than simply useful for humanity, in which nature's beauty, integrity, and diversity were considered to be inherently valuable. Whereas wise use had little patience for grizzly bears, preservationism almost venerated them.[5]

This philosophical divide is particularly contentious in the West. Because the federal government owns much of the land, it has considerable say over which environmental principles dictate its use. The nineteenth-century Homestead Act bequeathed 160 acres to citizens willing to put the land to productive use. But these lots weren't sufficient for a profitable farm—not even 640 acres were. Many homesteads went belly up, and the land revolved back to the feds. Public leasing was a historical compromise to resolve the situation but remains a source of conflict between states and the national government. States can't collect taxes on federal land, although the feds now return a portion of the leasing revenues to them. These lands, however, are not necessarily managed in tune with local preferences. This discrepancy came into greater focus as federal agencies saw their missions shift from optimizing timber and beef production to "ecosystem management."[6]

The increasing emphasis of federal land agencies on preservation paralleled the transformation of Western economies. So-called Old West industries—ranching, timber, farming, and mining—have declined in importance. Today, farming is a small sliver of economic activity in states such as Montana, Utah, and Nevada. The outdoor recreation sector is now on par with agriculture, and it is growing several times faster than the national economy.[7]

Economic changes brought newcomers to the West. People arrived from the coasts to open and staff bike shops, rafting operations, and other forms of environmental tourism. Because they tended to be from urban areas and/or highly educated, they brought preservationist environmental beliefs along with them. Stark cultural divides were amplified by increasing inequality. As one Idaho resident put it, "You either have three jobs or three houses—nothing in between."[8] Such divisions underlay the resentment of Old Westerners. They feel not only at the mercy of federal regulations and market forces but also increasingly nullified by the growing economic and political power of New Westerners.

The proposal to reintroduce grizzlies had to contend with this contentious history. FWS had studied the prospect of returning the bears to the Bitterroots for years already. The timber industry, however, was still reeling from restrictions on Pacific Northwest logging to protect the spotted owl. Workers and landowners alike worried about what the impact of reintroducing a threatened ursine, and the associated restrictions, would be on their livelihoods. Nevertheless, representatives of the timber industry and a wildlife protection nonprofit managed to hammer out a compromise plan. They proposed introducing the bears as a "nonessential, experimental population," which meant that ranching, farming, mining, and tree felling could continue in the same area. Their plan would also empower a citizen management committee to steer the whole process.[9]

To New Westerners, the plan looked like selling out to the timber industry. "When you crawl into bed with the enemy, you become the enemy," argued one environmentalist. They preferred the "conservation biology alternative." This would give bears the full protections of an endangered species, enforce habitat restoration, and prohibit road building in a recovery zone that spanned half of the state of Idaho. And decision-making power would remain with FWS bureaucrats.[10]

Old Westerners, for their part, could not be dissuaded from the idea that grizzly bears posed a danger to people: "I don't care about statistics. I care about being eaten. Somewhere out there, a grizzly is waiting."[11] Idahoans immense fear of grizzlies mirrored the disproportionate anger that ranchers often direct at reintroduced wolves. Coyotes take down eight times as many cattle, but it is wolves that represent the threat to traditional Western livelihoods.[12] Yet, people's tolerance of risks is almost never about raw numbers. Hazards evoke dread insofar as they are perceived to be unfair, unjustly imposed, or artificially contrived. Idahoans weren't responding so much to the danger of ending up in the belly of a giant ursine but rather to the likelihood of being chewed up and spit out by a policy that handed off land-use decisions to a distant scientific bureaucracy.

In any case, politicians in Idaho and Montana maintained their fealty to Old West principles. Mistrust of environmental proposals involving the federal government was simply too politically entrenched. A staffer for Idaho Senator Larry Craig argued, "This is more than just an experimental introduction, it is bid for controlling the entire Western way of life."[13]

Ultimately, polarization doomed the plan to restore grizzly bears. To Old Westerners, the animal symbolized the threat to their world. It felt like an imposition. Liberal, urban interlopers were forcing their vision of nature upon everyone else. As historian Michael Dax observed, environmentalists continued to dogmatically insist that any plan needed to "follow the strictest tenets of environmental thought." The effort languished amid partisan gridlock until the newly elected George W. Bush administration withdrew federal support.[14]

This same mutual mistrust characterizes conservation efforts elsewhere in the world. Despite having a radically different history, an effort to restore a wetland on the former border between West and East Germany failed in the face of farmers' resistance. They feared losing control of their lands. Moreover, farmers mistrusted the restoration plan. They associated it with arrogant "professional greens" in Berlin, who wanted "to impose an ideology" on them.[15]

Conflict over APR is only the latest manifestation of a longer history of urban–rural divisiveness. The same symbolic battle can be witnessed between Old West and New West, the same conflicting visions of nature and contrasting economic fortunes. Agencies have starkly reduced the amount of grazing allowed on federal land. Ranchers face stiff global competition. Phillips County, Montana, has half the population that it did in 1920. APR has been able to buy up considerable acreage in a short period of time only by taking advantage of economic despair. To be fair, the organization does make the effort to be a good neighbor. APR leases lands to ranchers and even pays them to install wildlife-friendly fences and to refrain from shooting coyotes. Yet, to opponents, this is just "calculated . . . grandstanding," obscuring the effort to "retain control."[16] That said, cultural and economic differences by themselves do not explain the persistence and emotionality of the divide. Deeper sources of antagonism can be found in the politics of environmentalism.

THE BEST AVAILABLE SCIENCE

The 1970s was the high point of American environmental legislation. Laws to protect wilderness areas, to clean up America's air and water, and to require environmental impact statements for federally funded developments were all implemented in quick succession. In a showing of bipartisan environmental consensus almost unknown today, 390 representatives and 92 senators came together in 1973 to pass a new Endangered Species Act. As it was being signed by President Nixon, almost no one recognized how powerful or how divisive the law would become.[17]

That changed soon after the bill's passage. The Tellico Dam being built in Tennessee was already controversial, requiring the inundation of villages, the destruction of Cherokee archeological sites, and the transformation of a free-flowing fishery into an impounded reservoir. Then, the rare snail darter was discovered in the Little Tennessee River. Because the three-inch fish's life course requires strong currents, environmentalists could leverage the new ESA to delay the dam. Impoundment would undermine the ecological conditions for the fish's survival. The Supreme Court's 1978 decision in favor of protecting the snail darter revealed that the ESA's language limited political considerations. Rather, decisions needed to be made "solely on the basis of the best scientific and commercial data."[18]

Congress has since made changes to the ESA to try to limit its power, including an exemption for the Tellico Dam. Senators recognized the danger of the act being used cynically, not simply to protect species but also "to attack . . . any activity [environmentalists] think is undesirable" by working with scientists "to find an endangered species or subspecies on the vast areas . . . impacted."[19] In any case, it ended up that the snail darter wasn't actually endangered; the species was later found to be swimming elsewhere in the region.[20] Nevertheless, Americans continue to live with the consequences of an endangered species law that scientized conservation politics. We can see them in the case of a rare Coloradoan rodent.

Most of the Preble's meadow jumping mouse's nine-inch length is made up by a slender tail (figure 5.1). [21] The rodent is small, furry, and chubby, with black beady eyes and a small nose that wiggles when it nibbles on plants,

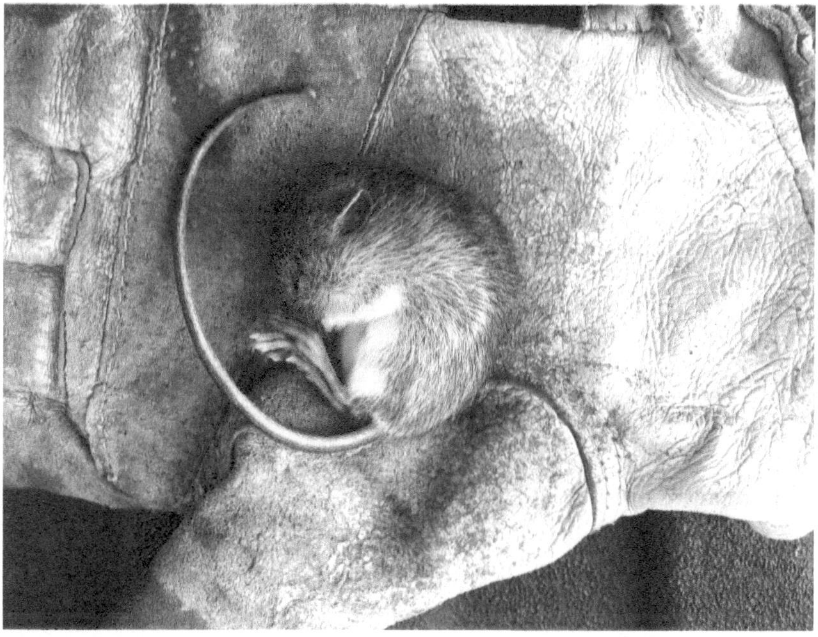

Figure 5.1
A Preble's meadow jumping mouse sleeping.

insects, and pollen. Despite their diminutive size, they can leap up to three feet to avoid danger, which I desperately hope is paired with a high-pitched squeak. In other words, the Preble's mouse is adorable. It crawls, climbs, and beds near the rivers and creeks of Colorado's front range, oblivious to the heated battle waged over its status as a threatened subspecies.[22]

The political stakes are as large as the Preble's mouse is small. The animal's ESA listing has led to some thirty-one thousand acres being labeled as "critical habitat." That designation prevents $100 million of development to ensure that a seemingly rare mouse has enough room to romp.

That listing, however, hinges on the mouse's scientific status as a bona fide subspecies. Defining a species is itself murky enough. Most of us learned in high school that species were groups of living things that could successfully breed with one another, but that's only one imperfect definition among several. Some scientists think that geographical isolation and visible anatomic differences are sufficient, whereas others look at the percentage of genetic divergence.[23]

What constitutes a subspecies is even less clear. As a reader might reasonably guess at this point, whether the Preble's mouse is really unique depends on which scientists one asks. And these researchers differ not only in the methods they use but also in their philosophical outlooks. Some scientists contend that the Coloradoan rodent is 99.5 percent similar to related mouse subspecies and is therefore not distinct.[24] Scientists on "team subspecies" take a more precautionary stance: "We . . . maximize evolutionary potential through the protection of populations on separate evolutionary trajectories . . . the most important aspect of biodiversity is protecting evolutionary potential."[25] In other words, these scientists believe that because the Preble's mouse was already geographically isolated, it should be protected. Even if it was not yet a clearly distinct subspecies, they reasoned that the animal was poised to *become* much more genetically unique over time. Which view represents the best available science?[26]

The endless taxonomic debate distracts from the underlying political question: Is it worth the cost to safeguard the existence of Preble's mice on Colorado's front range?[27] Unfortunately, the language of the ESA extends conservation protection only to species that are unique and under threat.

Even though many environmentalists would prefer to protect landscapes and habitats, the existing legal framework forces them to do that through trying to save endangered species. What is actually a dispute over land use becomes a polarized scientific debate about genetic isolation, morphological differences, and other tricky biological questions.

Philosophical differences between scientists very often make consensus elusive, if not impossible. Nevertheless, each side tries to make their side's incomplete research seem authoritative so that it can be leveraged into political power. The alluring promise of science-based policy, however, tends to undermine democratic collaboration. The plan to reintroduce grizzly bears was sunk, for instance, by environmentalists' rejection of citizen oversight as "nonscientific."[28]

The on-the-ground politics of the ESA aren't great. Species listing can have far-reaching impacts for landowners. FWS, for instance, declared one Louisiana timber field to be "essential for [the] conservation" of the dusky gopher frog. Turning the land into critical frog habitat would have cost the owner millions in lost property value and demanded costly interventions to maintain an open-canopy forest. The trouble was that the species hadn't been found in the state of Louisiana since 1965.[29]

The act's power is a double-edged sword. Fear of onerous and expensive land restrictions often leads to perverse outcomes. Many landowners encourage one another to "Shoot, Shovel, and Shut up" should a threatened species turn up on their property.[30] Indeed, some locals responded to the Preble's mouse's ESA listing by making their land less hospitable for mice, and most of them refused to have their property surveyed for the animal.[31]

Although APR is not being driven by the ESA, the scientization of conservation casts a shadow over the conflict. Environmentalists extol buffalos' scientific status as "ecosystem engineers' that can ostensibly fix much of what has gone wrong, ecologically speaking, on the Great Plains."[32] Skeptics cite their own experts to argue that buffalo are not necessarily easier on the land than cattle. And they question why APR's grazing plan wasn't subjected to a full environmental impact statement, suspecting (not unreasonably) that political concerns influence the process of discerning the best available science.[33] Insofar as both advocates and opponents tell themselves that they have environmental research on their side, they are talking too little about what matters to them. As a result, their collective ability to hammer out a common future on the prairie suffers.

LEGALISTIC BIOCENTRISM

"There's this incredible law where we can make people do whatever we want."[34] That's not exactly an accurate description of the ESA, but the effort to make reality fit that description has starkly polarized conservation politics.

American environmentalism once had widespread support. Idaho boasted a Democratic senator, Frank Church, for more than two decades, and he was at the forefront of creating wilderness areas. Archconservative Barry Goldwater was a member of the Sierra Club. The transformation of environmental protection into a focal point of America's culture wars was gradual, driven, in part, by the end of the broadly shared economic prosperity of the 1960s. In that context, new federal environmental legislation seemed to put preservationism and old-school conservationism into a zero-sum battle.

The tactics of mainstream ENGOs only became less democratic as conservative opposition turned fanatical. Self-styled eco-warriors rejected engagement with state governments and local people. Instead, they tried to push their agendas through national-level lobbying networks and to transform political disagreements into narrowly scientific and legalistic ones. They leveraged donations into endless lawsuits with the stated aim of holding federal agencies accountable to the "best available science" and the letter of the law.[35]

Overpowered litigation was inadvertently built into environmental legislation. The ESA and similar acts have "fee-shifting" provisions. These reimburse the legal fees of individuals and groups who successfully sue the federal government for failing to enforce the law. The strategy is made even more enticing by the Equal Access to Justice Act. This law was originally meant to help small businesses, social security recipients, and veterans to afford to sue the government in cases of bureaucratic misconduct. Legal reimbursement was supposed to be restricted to people of modest means. But a last-minute amendment removed that limitation for nonprofit organizations.[36]

As legal scholar Lowell Baier argues, the combination of fee shifting, the complexity of environmental science, and the rigid requirements of many environmental laws enable ENGOs, who are often already flush with donations, to wage legal battles at essentially zero cost. Simple administrative failures, such as when FWS fails to meet the twelve-month deadline stipulated in the ESA, often bring expensive lawsuits from the Center for Biodiversity (CBD) and a five-figure payout.[37]

Many environmental laws allow ENGOs to litigate whenever an economic activity or public decision involves a federal agency. For instance, we saw in the last chapter how the renewal of grazing permits at Point Reyes opened the door for the CBD to intervene. As a result of this legal framework and fee shifting, FWS and similar agencies spend a considerable amount of time and resources on litigation. They face hundreds of lawsuits each year, which can put them on the hook for hundreds of millions of dollars in legal fees.[38]

ENGOs contend that the arrangement ensures that federal environmental agencies do their jobs. They point out how endangered species listings lagged during the Reagan administration. Indeed, it took litigation for FWS to end delays in designating critical habitats for a rare Texas salamander. And the CBD leads an effort to force Maine lobster fishermen to finally experiment with whale-friendly equipment.[39]

Opponents, however, could argue that FWS funding gaps simply correspond to the publics' relatively low prioritization of conservation issues. And perhaps the slowness of listing species and critical habitats is mainly a response to difficult political realities, which are only worsened when government bureaucrats strictly follow the ESA's stipulations. In any case, the fact that the ESA only seems to "work" in response to endless litigation signals that it is not a well-designed law.

Whatever its merits for getting conservation work done, an exclusive focus on litigation has democratic downsides. ENGOs hold a legal trump card. They can almost unilaterally delay economic activities or developments that they don't like. This often includes grazing, residential development, or the delisting of species that are no longer endangered. Overwhelming FWS and other agencies with lawsuits is often an explicit strategy to ensure that ENGOs hold considerable power over environmental rulemaking. The CBD's long-serving director Kierán Suckling once likened the constant stream of new legal injunctions to "psychological warfare." His explicit goal was to make agency workers "feel like their careers are being mocked and destroyed." When the CBD halts grazing, shuts down logging, or prevents a dam from being filled, power shifts to them. According to Suckling, his organization is then in the "position of being able to powerfully negotiate the terms of releasing the injunction" and officials "become much more willing to play by [his] rules."[40] ENGOs leverage their power even more effectively when agencies settle in order to avoid litigation costs. It lets them shape agency rulemaking without the input of other stakeholders. ENGOs' legal strategy gives them disproportionate influence on the priorities of FWS and other agencies.[41]

To developers, ranchers, and others, this feels like bullying. No doubt they can and do use the same laws to fight bureaucratic overreach. Nonetheless, responding to environmental litigation is steadily becoming part of Western life. The CBD has blocked construction of high schools, dealt the fatal blow to the Southwest's timber industry, and intervened in innumerable cases involving grazing and water permits. Sometimes, the interventions seem frivolous. The CBD challenged rancher/investment banker Jim Chilton's routine grazing permit application, claiming that his land was overgrazed. However, the group's evidence gathering was shoddy, and Chilton was awarded more than $500,000 in damages in an ensuing libel suit.[42] Yet, some ENGOs appear so emboldened by the situation that they go out of their way to antagonize their opponents. Publications and campaigns from one Idaho ENGO, the Western Watersheds Project, routinely refer to ranchers as "welfare queens."[43]

The ESA has polarized conservation, making environmental compromises more difficult to achieve. The Bitterroot grizzly bear reintroduction plan faltered because ENGOs too fanatically insisted on following their understanding of the letter of the law. Any concession was seen as unthinkable. Some conservationists now explicitly try to work outside of the ESA's framework, although the implicit threat of listing likely helps motivate some of the cooperation. The National Audubon Society, for instance, has sought to protect the sage grouse via a mostly voluntary effort. Successes in building goodwill are imperiled, however, as Western Watersheds still threatens to sue FWS for not going far enough.[44]

Even though APR is being developed via voluntary land sales rather than litigation, the organization must contend with decades of built-up antagonism. Locals view environmentalists as deep-pocketed urbanites sending in high-priced lawyers to dictate Western land use. And, unfortunately, that isn't a totally inaccurate description of decades of conservation conflict.[45] Kierán Suckling and others have repeatedly confirmed rural residents' worst fears about environmental conservation efforts, especially by saying things such as "Ranching should end. Good riddance."[46] Many ENGOs think grazing should be halted, as one journalist put it, "even if it means that every cowboy in the West has to trade in his Wranglers for a McDonald's uniform."[47] It was naive of Gerrity and Freese to think that APR could reshape the Montana prairie without addressing decades of bad blood.

GOVERNING BY NUMBERS

Activists are being arrested by the dozens. They have blocked streets, skirmished with police, and forced open the front doors of government buildings. However, these aren't far-right protestors, anarchists, or environmentalists calling for an end to fossil fuels. They're Dutch farmers.[48]

Fanatical legalism increasingly characterizes environmental politics around the world, not just the United States. In the Netherlands, activists have found an alluring target for their lawsuits. The European Union is a powerful supranational bureaucracy, one that can compel individual nations to adhere ever more closely to continent-wide environmental regulations. It can force governments to make hard decisions that they otherwise would put off. But large-scale bureaucracy generally works through metrics, models, and indexes, not with people. And it ends up that usurping political decisions with mathematics often backfires.

Despite its diminutive size, the Netherlands is a farming powerhouse. It ranks only behind the USA in agricultural exports.[49] This achievement, however, is built upon some of the highest farm chemical use on the continent.

EU rule changes, however, have forced the nation to get a handle on its sky-high nitrogen levels. Both livestock waste and fertilizer applications emit nitrogen, which has a major impact on Dutch nature reserves. By the 2010s, nitrogen emissions had already fallen to half to what they were thirty years prior, but it wasn't enough. At the behest of the government, a model was developed that estimated the amount of nitrogen that researchers believed each hectare of land could tolerate. The government's environmental impact predictions would determine the level of permitting required for farming, building homes, and industrial activities.[50]

Enter Johan Vollenbroek, the head of Mobilisation for the Environment. Perhaps taking a cue from American ENGOs, he has repeatedly sued the federal government in order to force them to follow environmental regulations more strictly. In 2016, he sought to block construction permits for farms bordering two Dutch nature reserves. The country's high court ruled in Vollenbroek's favor: The government nitrogen licensing system didn't comply with EU rules. Thousands of farmers instantly lost their nitrogen rights. Other sectors have also felt the effects. A housing shortage grows, and in some areas, the electric grid reportedly can't accommodate new businesses for lack of permits.[51]

Vollenbroek's injunctions increasingly target individual farmers, many of whom have continued to operate without a license. Agriculture is an increasingly stressful way to make a living as farmers are saddled by economic competition, ever more stringent regulations, and legal action. Suicide rates have dramatically increased.[52] The *Stikstofcrisis* (nitrogen crisis) transformed farmers into speed bumps on the road to environmental progress. Their farms turn into mere debits on the national nitrogen ledger. As protecting nature becomes synonymous with improving the measures fed into large-scale environmental models, the Netherland's nitrogen crisis may be a sign of what's to come.

All modern government no doubt relies on scientific analysis and planning. Try to build highway systems or manage groundwater resources without it. Nevertheless, meeting most people's on-the-ground needs depends on tempering rationalistic measures, scenarios, and projections with a concern for the parts of life not easily represented in scientific models. For instance, traffic engineers have long optimized North America's roads for speed, for moving vehicles efficiently.[53] People who want to cycle to work, to safely walk their five-year-old to school, or to simply enjoy a coffee at a café without traffic noise aren't represented in the models and therefore lose out.

Rationalistic planning fails spectacularly when the model becomes everything, when quantitative measures and targets start hindering wise decision-making rather than helping it. The most egregious examples come from authoritarian nations, such as when Stalin forcibly mechanized and collectivized the Soviet Union's rural lands. Farmers were reduced to abstract economic units, whose sole purpose was to increase agricultural production. State agronomists unilaterally assigned crops to different parcels and jobs and quotas to locals, but they understood neither peasant society nor what was practical. Millions died in the ensuing famines.[54] There's no doubt that the stakes are not so high in the Netherlands. Yet, a similarly pathological politics has emerged.

The nitrogen model enables Dutch bureaucrats to prove that they are adequately regulating farmland emissions. Even so, it is an oversimplifying map of the human landscape. The model transforms electricity, farms, apartments, and workplaces into so many grams of nitrogen deposited per hectare. There is far more at stake than just getting the state's nitrogen spreadsheets to balance, most importantly people's livelihoods. But Dutch politicians' hands are tied by rigid targets, EU policy, and court orders. Governing by computer model stymies consideration of the messy political realities of the Netherland's countryside. The reaction has been explosive. Large-scale protests were just the start. A populist farmers' movement, the *BoerBurgerBeweging*, is now one of the country's most powerful political parties. They recently placed first in nearly half of the country's provincial elections.[55]

Anti-environmentalist populism, in fact, seems poised to grip the entire continent.[56] The Nature Restoration Law is currently making its way through the European Parliament, which would require countries to establish environmental recovery measures for 20 percent of their land and sea area by 2030. Requirements may even balloon to 80 percent by 2050. My conservationist friends are celebrating, but I worry about decades of divisive gridlock and lawsuits. Organizations representing farmers and fishers call the law "unrealistic" and a threat to rural livelihoods. An open letter from EU scientists, on the other hand, argues that opposition is "unjustified," that ecological damage and climate change are ultimately what threatens agriculture and food security.[57] Both groups are correct to some degree. The law is likely to push marginal farms and fishing operations over the edge, whereas protecting pollinators and fisheries will benefit the producers that do survive. Like for most polarized debates, the two sides largely talk past one another. The high stakes and urgency of the moment threatens to hinder sensible politics.

The man at the center of the Dutch nitrogen crisis, Johan Vollenbroek, certainly seems insensible. Pressed to comment on the human consequences of his lawsuits, he blames the Dutch government for its regulatory timidness, Christianity for imparting a "disastrous attitude" regarding nature, and farmers themselves for failing to listen to the science. He likens humanity to a pesky fungal infection but views himself as leading "Mother Nature's immune response."[58] That Vollenbroek sounds like an eco-warrior stereotype mixed with a Dutch-accented Bond villain would be hilarious were the human consequences not so high. He appears oblivious to his own callousness. Vollenbroek doesn't notice that such insensitivity, when translated into policy, feeds the growing fanaticism of his opponents.

Without alternatives to inflexible target-based legislation, other nations may end up following the Netherlands' example. Were the United States to follow the EU's restoration law, APR's efforts would look like a model for hitting mandated environmental goals. Those living in biodiversity priority areas or on landscapes considered low-hanging fruit for conservation conversion would feel the squeeze. Like Dutch farmers, they would become politically beholden to making ecological models turn from red to green. If we think environmentalism is polarizing now, we ain't seen nothing yet.

PEOPLE AS CONSERVATION PROBLEMS

Martin Saning'o was fluent in both English and the vocabulary of conservation. That much had become mandatory for leaders of the Maasai, a group of traditional pastoralists in Kenya and Tanzania (figure 5.2).[59] Yet, when speaking to an international audience of environmentalists and scientists in 2004, he told them, "You have made us the enemies of conservation." A hundred thousand Maasai had been displaced from their homeland, many in the name of protecting nature.[60]

Conservation is contentious because science can't actually tell us how humans should dwell on the earth. For all the marshaling of research data, models, and estimates of the value of ecosystem services, the debate is ultimately about which people are allowed to use the land and who gets to make that decision. The historical practice of conservation wasn't admirable in this regard, at least from contemporary moral standpoints. The experience of the Maasai has been tragically common.

Miwok, Paiute, and others Native Americans were first cleared from what is now Yosemite National Park in order to make way for white settlers. Early environmentalists then sought to preserve the area for nature-based leisure. The era's romantic movement held that time spent in nature would cleanse one's soul of the damages caused by "overcivilized" urban life. Many romantics had as low of an opinion of Yosemite's original inhabitants as the white settlers did. Early environmentalist John Muir considered the awesome cliffs and forest valley to be unsuitable for "such debased fellow beings." He believed that Native Americans failed to impart the "solemn calm" that he associated with wilderness. Eventually, Indigenous populations were fully evicted from the valley, as they have been from national parks across the globe.[61]

This process of setting up fortresses of nature by forcibly removing local peoples hasn't been limited to Indigenous groups. When New York made the Adirondacks into a state park, local white people and Native Americans alike found out that their efforts to live off the land were now illegal. Hunting was restricted to particular seasons, and certain techniques were prohibited, similarly for fishing and tree felling. What used to be common land became a patchwork of state and private ownership. The government began the difficult process of sorting out land rights, which turned many of the area's longtime inhabitants into squatters. State officials not only were motivated by the desire to safeguard opportunities for city dwellers to unwind in a pristine environment but also feared that "backward" locals, who lacked the environmental management credentials of state bureaucrats, would fail to properly steward nature.[62]

For the Black farmers of Pembrooke, Illinois, the process has been softer, but the eventual outcome may be no different. TNC and white suburbanites, working with FWS, have been buying up distressed farms at auction and putting them under conservation easement. Preserving the sandy black oak savanna landscape, the new owners believe, requires outlawing local traditions. They prohibit hunting, horseback riding, and picking wild fruit, but they allow hiking and birding. Locals feel insulted by the implication that they cannot care for the land. And they worry that if TNC gets its way, their community will eventually disappear in the name of conservation.[63]

Many ruralites feel that conservation is something done *to* them rather than *with* them. Even when softer approaches are taken, the reality is that protections are sought by wealthier urbanites and citizens of rich nations while poorer rural people pay the resulting costs. Much of the world's biodiversity is found within economically precarious nations in Africa and South America. This has allowed their governments to negotiate "debt for nature swaps," in which foreign debts can be forgiven in exchange for setting up conservation protection programs.[64] But these deals seem less like win-wins and more like selling your refrigerator in order to settle with a collections agency. They aren't exactly equal negotiations. The history of conservation is filled with morally dubious agreements. Much of the progress in protecting nature has been made in weak, highly unequal democracies, places where rural people are less able to fight eviction.[65]

Frustrated locals have often retaliated by lashing out at nature. Poaching becomes a way to get revenge at the government and distant environmentalists. In the Adirondacks, some residents purposefully targeted reintroduced elk and moose and set fires, as the forest came to symbolize their disempowerment. Others turned to poaching out of desperation because displacement undermined their ability to support themselves.[66]

Adding insult to injury, displacement tends to be followed by more conservation-friendly activities that cater to economic elites. Tension over the existence of the Maasai in the Ngorongoro conservation area in Tanzania is tied not just to environmental factors but also, according to locals, to the desire to dedicate the land to luxury safari trips.[67] New York's government facilitated the growth of private parks owned by the Rockefellers and other elites within the Adirondacks because their interests were seen as "identical with those of the State."[68] American Prairie Reserve donors fly in by helicopter in order to overnight in deluxe yurts.[69]

The Tanzanian government argues that the Maasai and their cattle are overpopulated, threatening environmental degradation and undermining wildebeest reproduction as climate change dries up local water sources. The Maasai, in contrast, claim it is all a pretext for clearing them from Serengeti National Park in order to secure a game reserve for the Dubai royal family. There are even some reports of villages being burned down. However, officials disavow any desire to evict the Maasai. Instead, they insist that relocations are voluntary, although some locals appear to choose this route only after rangers seize their cattle. The government's offer consists of a home on two and half acres, complete with running water and electrical hookup. The catch is that moving entails mostly abandoning their traditional lifeways, becoming more Tanzanian and less Maasai.[70]

Figure 5.2
A Maasai pastoralist walks with his herd into the Ngorongoro Caldera.

The goals of rural populations and conservationists have long been incompatible. A pristine, wild environment seemingly has no place for people and agriculture. APR has made some attempts at reconciliation, such as leasing to cattle ranchers and paying them to implement conservation-friendly strategies. However, they inevitably understand ranchers and their livelihoods as biodiversity problems to be solved. "Success" for APR unavoidably means an environment where people are absent. As demands grow louder to set aside more of the earth as protected areas, it is only more important to figure out how to support biodiversity without making rural populations into conservation problems.

OTTERLY DISRUPTIVE

The gangs are fiercely territorial. "It's like *Game of Thrones*," as one local resident put it.[71] Violent confrontations occur regularly along Singapore's waterways. These bands of ruffians, however, didn't come to the bustling Asian metropolis to sell drugs or smuggle counterfeit goods but rather to dig dens and fish for tilapia. They are otters, after all (figure 5.3).[72]

Some ninety-odd river dogs once again call the world's most densely populated city home. They recolonized the city's waterways after city policies reduced water pollution and expanded green spaces. The adorable creatures, however, do have their detractors. People sometimes arrive home to find their beloved, and often very expensive, koi fish being devoured by an otter family. For one British tourist, cuteness turned into carnage. After being savagely nibbled twenty-six times in the lower body by a roving gang of otters, he told reporters that he had feared for his life. Although both kinds of occurrences seem relatively rare, they nevertheless show the disharmonies of living with nature.[73]

When we think of conservation, we typically imagine people retreating from natural spaces. It is easy to forget that environmental restoration simultaneously brings more wild species into human territory. Rural residents have historically felt the impacts of species reintroduction, but urban expansion and environmental cleanup efforts will increasingly force urbanites to experience the inconveniences of wildness.

Figure 5.3
A Singaporean otter feasts on a fish.

Carnivores present the biggest human–wildlife conflict. Although reintroduction of wolves and grizzlies evoke fears, human attacks are incredibly rare. Government statistics, moreover, usually show that coyotes are a bigger threat to livestock.[74] Even so, risk is never just about the numbers. The dangerousness of wolves is partly symbolic. To locals, the species represents the environmentalists who have imposed the animal on them.[75] To focus on statistics also overlooks the psychological dimensions. Studies find that farmers living in areas where livestock are lost to wolves have higher levels of psychological distress: anxiety, sleep difficulties, and stress. They are haunted by the experience of finding grievously wounded animals and often feel powerless to prevent wolf predation on their livestock.[76]

In India, however, tigers don't just take farm animals, they literally eat people. Because the tiger is the country's national animal, and due to Hindu beliefs discouraging violence against living beings, rural Indians are unusually tolerant of wildlife attacks. The country hosts most of the world's tigers, partly thanks to Indira Gandhi's 1971 hunting ban. But there have been costs. A dozen or so of the big cats every year realize that hairless and tailless, upright-walking monkeys make for easy prey. Government compensation for the deaths of family members and livestock is difficult to obtain and is meager at that. And there are strict conditions that must be met in order for a tiger to be classified as a "man-eater." Officials can be slow to act and ineffective. One tiger in Maharashtra killed thirteen people before finally meeting with a hunter's bullet. Patience has worn thin in some villages, resulting in tiger poaching and violent attacks on forest department workers.[77]

It might be easy for a (sub)urban-dwelling environmentalist to wave off these risks as the costs of protecting nature, but doing so is unwise. As Renee Askins, the Wolf Fund founder who fought to return the canids to Yellowstone, argues, "Laws don't protect wolves, people protect wolves."[78] Goodwill safeguards endangered species better than restrictive regulations. Resented laws offer a precarious form of protection and often fuel backlash. Reflecting the entrenched rancor of the "wolf wars," Idaho politicians responded to ESA delisting by expanding the hunting season for the animal and by declaring the southern part of the state to be a "wolf-free zone."[79]

The effort to soften the environmental impacts of urban development brings similar, albeit lower-stakes, problems to cities. In *Accidental Ecosystems*, environmental historian Peter Alagona surveys the surprising adaptability of wildlife: white-tailed deer populate America's suburbs, coyotes prowl Chicago's lakefront, and bats roost under Austin's bridges. Wild neighbors bring new challenges. This is familiar to anyone who tries to keep wily squirrels out of their backyard bird feeders. Sometimes, the conflict literally involves life and death: Pets must be kept indoors, especially at night.

Proximity to wildlife brings new potential for zoonotic diseases. That said, some biologists believe that more diverse ecosystems can mitigate these risks. Although it is tempting to blame booming deer populations for Lyme disease, lots only get overrun with disease-carrying rodents and shrews because there are too few predators. Likewise, bats can control mosquito populations, and coyotes keep feral cat populations from exploding.[80]

There are also cultural benefits to wildlife. As I mentioned in chapter 3, some Colombians have grown fond of the hippos swimming in the Magdalena River. Many Singaporeans are infatuated with otters, maintaining social media pages to track their movements and organizing "rescue missions" when baby river dogs get lost.[81] On the other hand, it is hard to imagine Old West farmers learning to love the wolf.

APR's bison aren't especially ripe for human–wildlife conflict, although the odd visitor to Yellowstone National Park does get gored every year when they venture too close to a buffalo. That said, fences do need to be beefed up to contain them. APR's critics also cite the risk of disease spread, but the operation maintains that its buffalo are tested and treated for illnesses that can infect cattle.[82] As with wolf reintroduction, the symbolism of the buffalo outpaces the animal's on-the-ground impacts. They represent a vision of a West without ranchers. They signify the onward march of preservationism in shaping Western land use. Human–wildlife conflicts are as much about people's politics as animal behavior.

CONSERVATION IS WHERE THE MOUTH IS

British journalist George Monbiot doesn't seem to have ever had an anti-farming opinion that he could keep to himself. "Lab-grown food will soon destroy farming—and save the planet" is the title of one of his *Guardian* articles. While his main focus is livestock, he sees a bright future in which even pancakes become a "farmfree food." Monbiot isn't content to merely imagine new environmental possibilities. He sees his vision for the future as an imperative, one demanded by the litany of ills caused by agriculture: water pollution, carbon emissions, reduced biodiversity, and so on. Monbiot warns of a "domino effect" leading to "Earth system collapse," unless we rapidly substitute farms with rewilded lands.[83]

For anyone sharing Monbiot's view of nature as inherently fragile, traditional farming can look like a needless luxury. Faced with an imperiled planet, it is tempting to say, "Alright, then. Get rid of the whole lot. Let them eat lentils and lab-fermented protein!" When the problem is reduced to raw numbers, that seems like the rational solution. As Monbiot points out, farmers are a single-digit percentage of many countries' populations, but they decide what happens to most of the landscape. Grass-fed beef is itself a small portion of meat production. Most of it now happens in industrial livestock operations.[84] Why shouldn't "we" urban dwellers just vote to put farmers out of business?

The issue isn't so simple. First, people value farming for reasons other than just food production. Farm subsidies appear untouchable because of a broadly shared attachment to human cultivated landscapes and to the farmers themselves. Although that might be dismissed as romantic, it is not any less idealistic than environmentalists' valuation of pristine wilderness or their desire to protect parcels of land they may never personally visit. Berry and apple picking is as popular a pastime in the American Northeast as backpacking. People enjoy communing with bucolic versions of agriculture just as much as they like to connect with the idyllic, but no less artificial, wilderness of a national park.

The same complexity characterizes what we eat. Although a small percentage of people do track their meals' carbon footprint, most of us treat food as more than a number. We eat to partake in culinary traditions, to remember moments spent with family and friends, or because it simply tastes good. And, most of all, many people value the ability to choose what they eat. The mere whiff of a possibility that a proposed "Green New Deal" could affect meat production set off claims that Democrats wanted to "outlaw" hamburgers. Promises by European politicians to work to reduce meat consumption were met with similar levels of outrage.[85]

The dominant environmental narrative tries to sweep these complexities to one side. Consider one biodiversity scientist's tweet that "the voice of science [indicates] that meat consumption isn't an issue between greens and beefatarians, but rather, between humans and Earth."[86] If only.

Although many of us are familiar with social media infographics highlighting the greater carbon footprint of beef compared to pea protein or edamame beans, the matter isn't quite so straightforward. Whether today's livestock herds are a net emitter of methane depends on knowing exactly how big prehuman herbivore populations actually were, and estimates vary considerably.[87] One recent study found the climate impact of a square kilometer of wild Serengeti is no different than that of a parcel managed by Maasai herders. If you eliminate cows, they end up being replaced by wildebeests and other animals that altogether emit the same amount of methane.[88] Although that study's results may not apply to the Montana prairie, it still raises questions about how much cattle *really* count against humanity's carbon budget.

For the stockyard operations that produce a majority of American beef, another story can be told. Researchers with the Breakthrough Institute point out that the United States' beef operations produce far less emissions per kilogram than those elsewhere. They argue that selective breeding, feed additives, and other technologies could slash those numbers by half and considerably more so around the world.[89]

Finally, the choice of measure often biases discussion about the carbon footprint of food. Animal-based foods look bad when we measure carbon emissions per calorie or gram of food. However, human nutrition isn't just measured in energy or protein content. Proteins can vary in quality. Nutritious diets require vitamins and minerals.

One set of researchers took a different approach, evaluating different foods' emission while accounting for the level of essential micronutrients they provide: vitamin A, folate, B12, and so forth. According to the resulting metric, pork has the same climate impact as rice. Cow's milk production warms the planet no more than soy milk. Eggs have fewer emissions than tofu.[90] It is difficult to survey all these complexities and still conclude that the voice of science unambiguously says don't eat animal products.

Yet, concerns about grazing aren't limited to just carbon emissions. Livestock unavoidably take up space from wild animals, either by their grazing or by the need to cultivate feed crops. This is especially true in Brazil, where rainforest is cut down to make way for beef production. On the other hand, the impacts are likely less stark for areas that are already grasslands.

It seems reasonable that native herbivores would contribute to a more biodiverse and vibrant landscape. The scientific literature portrays bison as ecosystem engineers that fit better into the ecology of the Great Plains. They are better adapted to heat and drought, leaving the shores of streams alone until it's really hot. Local birds use bison hair to line their nests. A three-decades-long experiment found that buffalo grazing increased the richness of native plants compared with cattle, resulting in a more drought-tolerant prairie.[91] However, that only brings us back to the question of chapter 3: Which mixture of species deserve to live on a landscape? Why should we seek to exclude species that have better allied themselves with humans?

That question is central to the dispute. Advocates of "land-sparing" food production, such as Monbiot, argue that humans should live off of the smallest land area possible, leaving the rest of the environment for wilderness. The "land-sharing" approach argues for strategies that enhance the biodiversity of farmland and other human settlements, finding ways to include niches for wild species. Team sparing contends that sharing-based farming is inefficient, putting more land under the plow than necessary. On the other hand, sharers argue that intensive agriculture doesn't actually spare land, as farmers and firms invariably open up more pastures in order to grow for export markets and biofuels. Agriculture policy disagreements aside, the debate is, as journalist Fred Pearce put it, "about the kind of environment we want to live in."[92]

This same dispute is playing out for APR. Advocates of buffalo restoration view the landscape as best spared for native prairie species. Their end goal is incompatible with compromise scenarios, wherein farmers graze their livestock in ways that are friendlier to wild species or raise cattle–bison hybrids—or perhaps even a more diverse set of herbivores. If the trends enabling APR's land purchases continue, there may be little question of what kind of relationship with agriculture will win out. Economic forces will lead to a land-sparing system, at least in the American West. Meat will come from intensive industrial livestock operations and perhaps laboratories.

George Monbiot's advice to farmers is "get out now." Journalist Anna Jones contends that Monbiot's extreme rhetoric might just be a kind of showmanship, an intentional provocation meant to shock people into action.[93] Even so, such rhetoric antagonizes farmers and their allies. It gets under people's skin, questionably leveraging science to claim that *there is no alternative* to a radically transformed food system. Whatever good Monbiot's shtick has done for his book sales, it doubtfully earns his side many friends among agriculture workers or meat eaters. It is symbolic of what has gone wrong with environmentalism in general.

TRANSFORMATIVE CHANGE

"Reverting human impact on biodiversity would require . . . a fundamental, system-wide reorganization across technological, economic and social factors, including paradigms, goals and values."[94] The values report of the Intergovernmental Science-Policy Platform on Biodiversity and Ecosystem Services (IPBES) insists that only "transformative change" can save the planet.

The idea of transformative change has a long history. It is practically a cliché to argue that the scale and urgency of environmental problems imposes radical demands upon humanity. George Perkins Marsh, for instance, justified expanding parklands with the argument that "the earth is fast becoming an unfit home for its noblest inhabitant, and another era of equal human crime and human improvidence . . . would reduce it to such a condition of impoverished productiveness, of shattered surface, of climatic excess, as to threaten the depravation, barbarism, and perhaps even extinction of the species."[95] What sounds like an environmentalist's lament from the hippie era was actually written in 1864. A century later, Paul Ehrlich and other ecologists argued that avoiding certain catastrophe requires sharply constraining people's ability to have children.[96] Today's climate activists, such as Naomi Klein, claim that saving the planet requires not just embracing new technologies and ways of living but also reorienting the whole economic system away from capitalism.[97]

Transformative change is, of course, one legitimate environmental perspective among many. However, equating the saving of nature with society-wide transformation is poised to make environmentalism only more contentious, for it sets sharp limits on our political imagination.

The IPBES values report, for instance, speaks of including diverse values, but it is evident that some values are best left behind. The report's authors argue for leveraging "governance systems" in order to reduce the values of individualism and materialism. This requires "changing paradigms about how to pursue a meaningful life" toward those "aligned with more respectful relationships among humans and towards nature."[98] This sounds a lot like the "new kind of [ecological] people" that Aldo Leopold hoped for. The fact that the authors preemptively blame human "cognitive limitations," should people resist transformative socio-ecological change, doesn't help the perception that it is a plan to be implemented *on* citizens rather than *with* them.

The report also understates the radicalness of large-scale social engineering for the environment. Pathways such as "degrowth," which proposes that citizens in highly developed nations scale back their standard of living, are highlighted with little acknowledgment of how contentious they are with the broader public.[99]

Probable infeasibility isn't the report's biggest problem. Most troublesome is its political partiality, a sentiment echoed by many of the biodiversity scientists that I have talked to. Conservative readers would find it suspiciously convenient that ostensibly scientific plans for living in harmony with nature more or less correspond to the policy preferences of left-leaning environmentalists. It doesn't help that activists such as Klein say as much for climate change, that it "supercharges the pre-existing case for virtually every progressive demand on the books, binding them into a coherent agenda based on a clear scientific imperative."[100]

Environmentalism has been long plagued by the belief that deeply political disputes can be transformed into scientific issues. However, by turning nature problems into matters solved by expert administration rather than via democratic negotiation, environmentalists help foment populist opposition to their own policy goals.[101] Cases such as the Dutch nitrogen crisis demonstrate as much. To be fair, the IPBES values report's authors call for including the nature values of almost everyone, but they seem to already have a clear idea of where public inclusion is supposed to lead us.

Calls for system-wide transformations, moreover, have become increasingly totalizing. As German sociologist Sven Hillenkamp argues, activists more and more depict environmental progress as inseparable from leftist positions on capitalism, gender pronouns, police brutality, and veganism. Fridays for Future leaders demand that attendees not only agitate for climate policy but also fight against their own internal white-colonial privilege. No doubt the causes of environmental harm are myriad. But equating the protection of nature with a narrow set of partisan positions on capitalism, inequality, gender, and other hot button issues undermines coalition building.[102] In that respect, left-wing claims to the effect that conservatism is inherently anti-nature hasn't helped things either.[103] Environmentalism increasingly fails to function as a big tent movement. Instead, it invites only those already embracing progressive politics.

Apart from sneaking in left-wing values under the guise of science, demands for transformative change fail for an additional reason. They frame environmental solutions as necessarily *comprehensive*. Advocates of transformative change portray climate change and biodiversity loss as only solvable via radical social, economic, and political upheaval. This not turns an already difficult problem into a hopelessly herculean task but also needlessly amplifies the political stakes. Part of what drove the sometimes-violent response to COVID-19 pandemic controls was that they majorly disrupted people's lives. Jobs were lost. People were isolated. Citizens faced mandated testing, distancing, mask wearing, and vaccination. There are few transformative political changes throughout history that didn't foster animosity or even outright violence. Many people are understandably wary upon hearing of proposals for a "system-wide reorganization" of society to protect nature.

Old Westerners also rightly mistrust calls for transformative change. Environmentalism hasn't exactly been friendly to the people laboring in natural resource industries such as timber, oil, mining, or agriculture. Former chief scientist of TNC, Peter Kareiva, recalls watching loggers quietly sitting in the back row of a federal court and then hoisting their children onto their shoulders. They protested silently as lawyers and scientists determined whether their livelihoods and futures were less valuable than saving the spotted owl.[104] Although most rural residents say they value the environment and conserving species, few of them trust the federal government to fairly solve the problem.[105]

In contrast to demands for transformative change, environmental progress has typically been incremental and messy. As I mentioned in chapter 2, the United States' climate change–fighting Inflation Reduction Act was an amalgam of disparate political projects. Renewable energy has made its biggest inroads in "red" states as people there learned that green energy can be a moneymaker.[106]

Political progress entails hammering out compromises, finding unexpected allies, and making sometimes-ugly trade-offs. Coalitions can't be built without an effort to help people learn to trust that ensuing changes won't harm them. By laundering values behind the facade of science, by making environmentalism seem exclusive to leftists, and by portraying high-stakes societal upheaval as the only pathway toward progress, the rhetoric of transformative change undermines the goodwill needed to collectively respond to environmental problems.

MENDING FENCES?

Goodwill doesn't seem to be on the horizon in Montana. Governor Gianforte and the state's Stockgrowers Association are waging a legal battle to reverse Bureau of Land Management rule changes favoring APR. These changes allow bison to graze year-round on federal land and permit the organization to take down fences. Although there may be a scientific argument for relaxing BLM's normally strict requirements for buffalo, the cattlemen only see unfairness. What about the decades of efforts they themselves have made to restore historically overgrazed prairie land and to allow an albeit limited return of wildlife?[107]

APR has cultivated stronger relationships with local Aaniiih and Nakoda tribal members, inviting them to watch bison releases and donating buffalo meat to the reservation. Even if the effort is led by white people, the return of the historical food source for the Aaniiih and Nakoda people symbolizes a reversal of centuries of repression of their traditional lifeways. This new coalition will work so long as those lifeways align with the environmental vision of APR's advocates and donors.[108] However, it further complicates the effort's politics, calling to mind the historical conflicts between the region's Native Americans and white farmers.

Ranchers' reaction to APR seems disproportionate at first, but the region's history uncovers the roots of resentment. Rural residents often feel excluded from economic and cultural shifts happening around them. Environmental policy can seem to unfairly empower urban environmentalists and their lawyers. Expert-driven politics frames rural residents' concerns as nonscientific and unecological and thus unworthy. Even though environmental regulations have been necessary to curb overgrazing, to rein in logging practices, and to clean up the air and water, only ruralites had to deal with multiplying wildlife conflicts, shrinking profit margins, the increasing threat of eviction or land-use restrictions, and the other consequences of conservation. They have had to bear these burdens while simultaneously being derided as obstacles to progress. When they look at environmentalists' visions for the future, rural people too often see a world that leaves them behind. As urban populations swell, there's the risk that they will simply be outvoted by urbanites enamored with the idea of wilderness—as has happened in Colorado's decision to reintroduce wolves.[109] But democracy is more than mere voting; it is a system where people have influence over the decisions that affect their lives. A democratic approach to conservation requires us to do better.

Even for those of us who don't live off of the land, biodiversity policies will increasingly impact our lives. They will raise the prices of food and other goods and services, and restrict where people are allowed to camp, hunt, fish, ski, drive off-road, and build a home. Clashes between humans and wildlife will become more common, including in urban areas.

Conservation goals will come into conflict with other environmental objectives. ENGOs have recently pivoted to confront a new enemy: renewable energy. Lithium mines, solar installations, wind turbines, and transmission lines eat up a lot of land. In the United States, environmental groups see fragile arid ecosystems, which boast rare wildflowers, desert tortoises, sage grouse, and other unique species, as under threat. Indeed, the CBD has recently petitioned the FWS to list a unique Nevada butterfly, a move that will likely block a proposed geothermal energy project.[110]

Even the return of wildlife may have climate consequences. The resurgence of beaver populations adds around a million tons of additional methane to the atmosphere each year, nearly as much as from US oil production. The mandate to combat both climate change and biodiversity loss will pit environmentalist against environmentalist.[111]

If more radical environmental coalitions have their way, citizens in developed nations will be asked to reduce their standard of living while residents of still developing countries will be encouraged to lower their expectations for the future. I doubt that such a vision aligns with what a lot people want or what they want for their children. Dreams of harmonious ecological futures will be a hard sell outside of committed environmentalists, no matter how detailed and well founded the models are or how urgent the pleas from some scientists may sound. Learning to take the fire out of conservation disputes in rural areas will be much-needed practice for the bigger and thornier political battles that await us.

While reading the United Nations Montreal agreement and watching the Nature Restoration Law work its way through the EU parliament, I can't help but see these policymakers as akin to their 1970s American counterparts. They act in horror over what seems to be a wanton destruction of nature and seek to combat environmental damage with the same level of zeal. But like the senators who signed off on the ESA, they fail to imagine that the transformative changes they seek may end up backfiring, that calls for radical action may actually exacerbate the underlying conflicts and make biodiversity targets harder to reach.

The dominant mode of environmental politics since the 1970s long ago reached its breaking point. It has been too technocratic, antagonistic, and legalistic to bridge the political chasms dividing people with a stake in conservation decisions. If conflicts are to be productively resolved, we will need better ways of reconciling competing ideas about nature and the proper means and ends of environmentalism. We need biodiversity democracy.

6 RECONCILIATION

Most Americans and Europeans know of elephants only as charismatic megafauna. The giants are majestic creatures that fill us with amazement. We watch nature documentaries and flock to zoos just to catch a glimpse of them. For the people who live alongside the world's largest land animal, however, they can be a real pain in the ass.

Hundreds of unfortunate people are killed by elephants every year. The pachyderms regularly raid croplands in order to satisfy their dietary needs: some 350 pounds of plant matter daily. Entire fields can be decimated in a single night, an outcome that poor farmers can scarcely afford.

The consequences of the West's love of megafauna aren't lost on Botswana's president. The nation has seen their elephant population boom to more than a hundred thousand animals. After the German government moved to deter their citizens from trophy hunting, an incensed President Masisi threatened to send tens of thousands of elephants north, so that the Germans could "live together with the animals, in the way you are trying to tell [Botswanans] to."[1]

After noticing that elephants avoided foraging on trees with bees in them, zoologist Lucy King had an idea. Maybe beehives could be used to deter the animals from farmers' fields. Her design involved setting hives on fence poles and coupling the bee colonies together with wires. King hypothesized that intruding elephants would pull on the wires, which would jiggle the hives and trigger angry swarms of soldier bees. The buzzing pollinators, in turn, would sting the elephants in their sensitive eyes, ears, and trunks.

As studies and anecdotal experience confirmed the relative effectiveness of beehive fences, their usage has spread from Kenya to India.[2]

Bee fences can be expensive to build. Some estimates put the cost at more than $1,000 per acre. Fortunately, conservation organizations will often foot the bill, and keeping bees brings additional benefits. They pollinate crops. Farmers, moreover, can diversify their income stream by collecting and selling honey. That said, the fences aren't a foolproof solution. Bees sometimes abandon their homes. Some elephants are tenacious enough to break through. Local honey badgers, as their name suggests, will sometimes eat the hives. Elephant conservationists are investigating additional solutions such as chili oil, noise makers, and digital technologies to track elephant herds.[3]

Beehive fences remain a work in progress. Nevertheless, they show the potential of humane alternatives to traditional strategies of elephant deterrence, which sometimes included retaliatory killings and leaving traps of explosive-laden fruit. Biologist Michael Rosenzweig would see beehive fences as a kind of "reconciliation ecology." Rosenzweig defines this as "the science of inventing, establishing, and maintaining new habitats to conserve species diversity in the places where people live, work, or play."[4] Bee fences help reconcile elephant conservation with humans' need to grow food.

Rosenzweig's 2003 book, *Win-Win Ecology*, illustrates how the presumed conflict between protecting and using the environment has been cleverly, and sometimes accidently, reconciled.[5] Examples include an underground Israeli restaurant, which offers patrons a spectacular view of an artificial ecosystem. The vibrant coral reef on the other side of the restaurant's windows were built up from rescued coral fragments, the result of a collaboration with a local research center. Land managers of Eglin Air Force Base cultivated longleaf pine trees and learned how to drill artificial nest cavities in them, providing habitat for the rare and fickle red-cockaded woodpecker. Bird lovers have learned how to design nest boxes that let bluebirds in but exclude non-native starlings. Cattle tanks and ponds can provide a lifeboat for Chiricahua leopard frogs. American crocodiles thrive in the cooling canals of a Florida power plant. Planting corridors of fig trees can help orangutans survive in rainforests taken over by palm oil plantations.[6]

Such examples are no doubt inspiring, the kind of thing your pathologically optimistic aunt might compulsively share on Facebook. Yet, they all have their downsides and trade-offs. Moreover, achieving reconciliation is as much a task for politics as it is for science. Repeating the happy accidents of crocodile-friendly canals and the innovative creations of beehive fences requires supportive legislation, rules, and institutions. What efforts exist to achieve conservation win-wins? How do they help temper environmental polarization? What are their limitations? What future is there for reconciliation biodiversity?

POST-NORMAL SCIENCE

Managers at CD Aero wanted to eliminate n-propyl bromide (nPB) from their capacitor production process. nPB had been used as a degreaser for years, but it came with costs. The associated machinery ate up a lot of space. Some five thousand pounds of the chemical leaked into the environment every year. Finally, its use made additional safety precautions and governmental oversight necessary—and for good reason. A 2013 exposé by *The New York Times* found that furniture factory workers chronically exposed to low levels of nPB suffered from numb hands and feet, nerve pain, and other ailments. Fortunately, CD Aero is located in Massachusetts, and they were able to work with a state agency to find another way forward. Scientists with the Toxics Use Reduction Institute (TURI) helped them design an alternative process that was not only less hazardous and environmentally harmful but also did the job. The substitute degreaser actually ended up simplifying the manufacturing process, saving the firm money.[7]

I'll forgive readers if, at first glance, they didn't exactly have their minds blown by that last example. But it is relatively uncommon for a business to voluntarily collaborate with government to detoxify the workplace. The main barrier is what passes for regulatory common sense: determine if a substance is toxic and then, if so, ban it. The Toxic Substance Control Act (TSCA), for instance, asks the Environmental Protection Agency (EPA) to assess the risk of hazardous chemicals and determine which ones pose an "unreasonable risk." Risk assessments are long and expensive scientific studies that try to weigh the relative harms and benefits of prohibiting a potentially toxic industrial chemical. However, those studies only get done after a compound has already become widely used, often at a point in time when it seems irreplaceable.[8]

Because of the high stakes of chemical bans, industries nearly always push back. Even an effort to restrict the making, distributing, and use of asbestos-based products in 1989 faltered in the face of lawsuits. The EPA has avoided bans in the years since, trying instead to negotiate voluntary phaseouts. The 2016 Lautenberg Act has helped only a little. The EPA now aims to have twenty risk evaluations ongoing at any given time. However, progress remains incredibly slow. The agency has to work through thousands of potentially hazardous chemicals that were "grandfathered" in under the TSCA.[9] It is tempting to think that regulation isn't "tough" enough, that giving the EPA more regulatory teeth will finally rein in chemical toxicity. That view is misguided.

Toxic chemical assessment is a form of "post-normal science." It exists in a space where the facts are uncertain, the stakes are high, values are in dispute, and decisions are urgent (figure 6.1).[10,11] Determining the risk of a chemical is no simple business. The science is complicated by questions about what level of exposure is realistic. How much does a typical consumer or worker actually ingest or breathe in? Moreover, it is often very challenging for animal or epidemiological research to produce authoritative, unassailable conclusions regarding toxicity. Lab rats aren't people, and comprehensive statistical data are hard to come by.

Even then, evaluating the risk is inherently subjective. To some people, any amount of risk, such as allowing chemicals in plastics that might affect our hormonal system, is irresponsible. Others point to the benefits of modern chemical products, such as in lifesaving medical equipment. Wrongly banning a substance would unnecessarily cost industries billions of dollars and lower people's quality of life, but failing to act can allow hazardous chemicals to become widespread in the environment and in human bodies.

These complications put toxic substance control in what I call the "zone of political inaction." It is a situation that usually can't be resolved by doing more research. Industry scientists easily find flaws in university or governmental toxicology studies. Even when science seems to be moving toward a clear consensus, the political situation only tends to worsen. When "tough policies" (i.e., bans) appear on the horizon, industries are only more strongly incentivized to fight regulation and nitpick the science. They amplify the uncertainty in the effort to bring the stakes back down again. Consider how tobacco companies successfully delayed restrictive legislation for decades by denying the highly probable link between smoking and lung cancer.[12] Focusing on trying to reduce scientific uncertainty rarely gets us back to "normal politics."

There are, however, success stories. In 1987, dozens of nations formally agreed in Montreal to phase out chemicals that appeared to damage the ozone layer. What many people forget is that ozone science didn't solidify until *after* the agreement. Shelves of books and articles have been written on the complex international relations that enabled the Montreal Protocol. Nevertheless, one key factor was that technological alternatives were already in development. Companies such as DuPont quickly recognized the potential profitability of becoming an early supplier of ozone-friendly chemicals. We didn't actually have to consider the possibility of living without refrigerants and other useful compounds. As science policy scholar Daniel Sarewitz puts it, the existence of alternatives "made it possible to meet the goals of multiple constituencies with conflicting values and worldviews."[13]

When faced with post-normal science, agreement or consensus about the risks is never going to happen. We don't get back to normal politics through brute force efforts to lower the perceived uncertainties. The all too typical result is getting trapped in a doom loop where the science never makes the risk seem *certain enough*. Instead, uncertainties naturally lessen as a by-product of lowering the political stakes. Once the path forward is made considerably less painful, becoming anything but existential, uncertain claims about risk become far easier to accept. DuPont's work on chlorofluorocarbon alternatives meant that an authoritative consensus within ozone research was no longer a precondition for action.

This is what makes organizations such as TURI so special. Rather than wage a high-stakes political battle to outlaw chemicals, the institute works to discover mutually beneficial solutions. They look for answers that partly

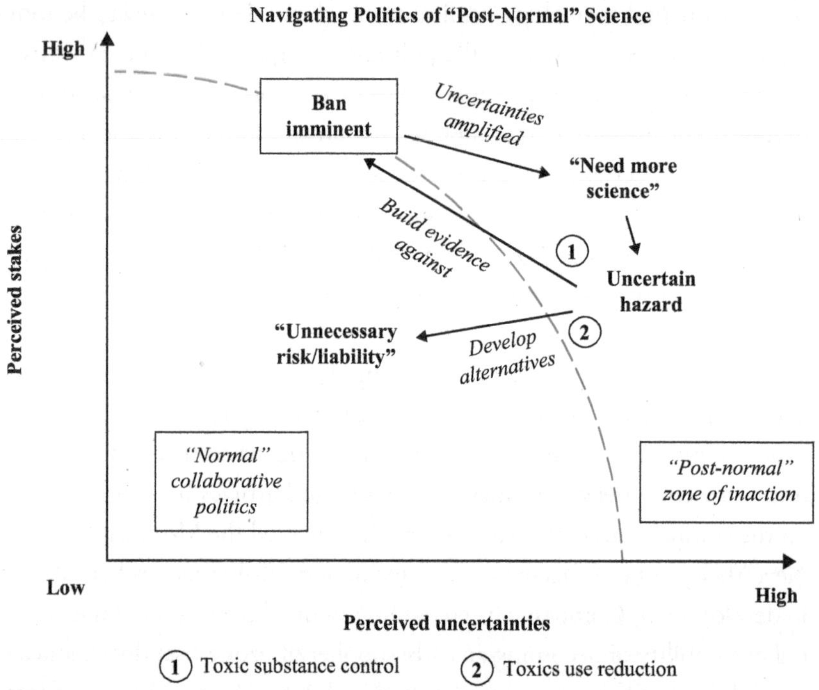

Figure 6.1
The contrast between toxic substance control approach and toxics use reduction pathways.

reconcile a company's business goals with public interests—how to stay profitable while also reducing the human and environmental harms of their production processes.[14] Most importantly, the organization provides a venue for hammering out compromises. Despite not directly involving governmental rule makers, voting, or the strict application of legislative statutes, TURI accomplishes a *more democratic* solution than the TSCA. The interests of people impacted by consequential chemical choices are represented in a negotiation that effectively solves a public problem. Since TURI was established, industrial usage and releases of known or suspected carcinogens has steeply dropped in the state of Massachusetts. This has not been the case for the US more broadly.[15]

TURI is also a bottom-up approach, one that involves businesses rather than subjecting them to a distant, faceless bureaucracy.[16] People often resent feeling excluded from governmental decisions, even for fairly mundane issues. A quick Google search uncovers videos of the vicious animosity present at a lot of city council meetings and public hearings, especially when local governments approve a new high-rise development or local homeless shelter. People want to feel like they can influence the decisions that affect their lives. It is infuriating when governments act as if people only need to be informed of what state administrators think is best for them. Environmental politics is no different.

As the last few chapters have shown, biodiversity conservation is post-normal science.[17] The stakes are high, values are in dispute, decisions seem urgent, and the science is uncertain. Achieving long-lasting success in reconciling biodiversity protections with humanity's other needs and desires will depend on our ability to create, maintain, and grow TURI-like conservation institutions. Moving from the polarizing zone of inaction to something more like normal politics means avoiding the high-stakes politics of catastrophism. It will require that we focus less on asking scientists to provide *answers* regarding environmental risks, facts to justify transformative policies, or consensus on how we ought to live with nature. Instead, we ought to emphasize collaborative *solutions* that reduce conservation hazards. What actions can take the fire out of environmental polarization and allow nearly everyone to feel as if they can beneficially take part in environmentalism?

COMPENSATING FOR DEVELOPMENT

These days, chimpanzees guard the river. They chase off villagers who come to gather water, bathe, or wash clothes. They raid local fields and threaten to spread disease. Conflict with the animals intensified in Guinea after the World Bank Group helped fund new bauxite mines. Even worse, streams have been polluted, farm fields have been contaminated with mine dust, and some locals feel that they have nothing left.[18]

Financing for the mine had strings attached—namely, the requirement to establish a "biodiversity offset." This came in the form of establishing the Moyen-Bafing National Park. The park compensates for the deaths of chimps near the mine by better protecting a different population of the animal located some two hundred miles away. However, this chimp community is now potentially threatened by a proposed dam, which would inundate a portion of the park.[19]

The use of biodiversity offsets is an example of the "new conservation" approach advocated by Peter Kareiva and Michelle Marvier, who were introduced in chapter 2. One of its central tenets is "working with, rather than against, resource users."[20] New conservationism is a kind of compromise. It recognizes that conservation can't be just about trying to preserve "fortresses" of pristine nature, that development can't always be stopped. People need resources and jobs, especially in impoverished nations.

That said, not letting the perfect become the enemy of the good isn't always an unalloyed boon to biodiversity. Or, better stated, the new conservation is fraught with its own challenges.

A kind of new conservationism actually predates Kareiva and Marvier's proposal. The Endangered Species Act, for instance, allows for so-called safe harbor agreements. These guarantee landowners that they will not be subject to additional restrictions if they voluntarily protect habitat for endangered species. Red-cockaded woodpeckers, for instance, nest in pine trees in the southeastern United States. One study found that locals logged more trees when woodpecker colonies were plentiful, likely in order to keep the species off their land. Participation in a safe harbor program reversed this pattern of habitat destruction. Nevertheless, landowners find it frustrating to negotiate an agreement with the US Fish and Wildlife Service due to its complexity and length. This limits the ability of the program to lessen the ESA's perverse incentives.[21]

At a larger scale, Clark County, where Las Vegas is located, worked out a sort of offset deal with FWS. They have sought to balance the city's growth with the protection of the desert tortoise, the species that sparked the armed resistance to BLM officers near Bunkerville. FWS permitted the city to relocate tortoises and develop some hundred thousand acres of federal land in the Las Vegas Valley. In exchange, the city agreed to establish habitat reserves, fund tortoise research and rehabilitation, and purchase and retire grazing permits, among other changes. Similar efforts in California have succeeded in offsetting every acre of development with new critical habitat areas. The desert tortoise now has considerably more conservation area to roam.[22]

The species, however, is a holdover from when the region was far wetter. The Mojave desert tortoise has continued to decline, despite millions of dollars being spent on conservation. The reptile is challenged by a boom in raven and coyote populations, a hotter climate, and disease outbreaks. And it struggles to survive being relocated. Although cooperative agreements may not be enough to save the species, they have at least succeeded in building up political experience with reconciling development with nature protections.[23]

There was, however, one big oversight. The outcome prioritized urban development over ranching. And facilitating Las Vegas's sprawl was estimated to have caused the deaths of hundreds of tortoises each year, an irony probably not lost on desert ranchers.[24] In any case, had urban interests not been so decisive in the negotiations, maybe we would have never heard of Cliven Bundy.

Realizing similar partial successes around the world isn't so straightforward, especially in impoverished, weak democracies. Nearly all Guineans live on less than $10 a day, and the West African nation ranks 147 out of 180 in terms of the perceived level of governmental corruption. Although the mining company insists that "resettlement . . . was done following a participatory and inclusive process," to the villagers, it was bait and switch. They told reporters that they were swindled through false promises of local development, jobs, and large payments for land.[25]

Offsets have been successful in some places. An often-cited exemplar comes from Madagascar, where the loss of forest cover from a large nickel mine was balanced by discouraging tree harvesting by farmers elsewhere. Still, one comprehensive review of offset programs questions their effectiveness. Its authors point out that "no net loss" in biodiversity was achieved only a third of the time. There's simply not enough people watching and making sure offsets aren't undermined. Nor are institutions in place to impose punishments for violations.[26]

For Guinea, the trouble is that the offset is going to be partly undone by a Chinese-financed dam. It is the pet project of the country's strongman president, who sees no need to offset the chimp populations whose habitat will get flooded. Still, on-the-ground practitioners are quick to point out that, for all their problems, biodiversity offsets are a substantial improvement over the previous status quo, which was doing "nothing."[27]

Those improvements can nevertheless result in perverse outcomes. Whether a new protected area actually compensates for development impacts depends on tricky estimates. Resource users have an incentive to exaggerate the conservation risks faced by species in the "offset" area in order to make the case for development easier. Moreover, lands that can be swapped for development projects eventually become scarce and lucrative commodities. In Australia, some people have opposed establishing protected areas because local residents would then no longer be able to sell the land as a biodiversity offset.[28]

Because the focus is almost exclusively on the environmental impacts, large social risks often get ignored. In the case of Guinea, villagers were relocated from a shaded valley to a sunbaked hilltop with poor soils. Others were moved to lands that weren't yet ready for farming. Despite having their traditional livelihoods disrupted by wildlife conflicts, dried-up streams, and mine pollution, the people most affected by development enjoyed few of its benefits.[29]

Even the more successful use of offsets in Madagascar brought new problems. New forest restrictions had to be enforced by villagers' neighbors (and sometimes the police). Efforts to help locals embrace new agricultural techniques and sources of livelihood failed to deliver. Many farmers had to take up odd jobs to survive.[30]

For all their problems, biodiversity offsets represent an incremental step in the right direction. Ultimately, they lessen the stakes of conservation conflict. That said, they need work. Most of the world's biodiversity hot spots and the sources of ore needed for green energy technologies are both located in impoverished, questionably democratic nations. Because rule of law tends to be weak in these places, achieving both environmental and social justice goals will be a challenge. Offsets can turn biodiversity into an abstract matter for bureaucratic administration, which corrupt political elites can take advantage of to ride roughshod over rural populations—all while claiming to act on behalf of conservation. Without democratic control, these efforts are bound to leave human suffering in their wake. As Ousmane Bah, a Guinean assistant village chief, put it, "We don't own anything [anymore]. We don't even own ourselves."[31]

PARTNERS IN BIODIVERSITY

Dutch farmers were originally skeptical of government efforts to get them to participate in programs for protecting nature. They changed their tune not because they were persuaded to become ecological true believers but rather by changing their relationship with the government—and each other.

So-called agri-environmental schemes pay agriculturalists to help conserve biodiversity by sowing flower seeds, waiting to mow their grass, leaving some grain unharvested to serve as bird food, and other strategies. But Dutch farmers found the financial incentives to be insufficient. The program was just too much hassle. They were individually responsible for arranging an agreement with the national government and satisfying a confusing mess of bureaucratic requirements. Even worse, administrative mistakes could result in financial penalties.[32]

The main outcome was mutual mistrust. Farmers felt that "those people in the Ministry" were ignorant about their circumstances. Government officials and members of nature organizations, in turn, were dismissive of farmers' knowledge, concluding that "they know nothing about biodiversity, [because] they are only educated to produce."[33]

Then, the new Agrarian Nature and Landscape Management policy changed how enviro-agricultural schemes were organized. Farmers now sat together with their neighbors to cooperatively develop and implement plans at a larger scale. This decreased the time spent with bureaucratic issues and fostered a collective identity around conservation activities. A program that was previously frustrating now instilled trust, built community, and fit biodiversity into farmers' already-existing desire to see themselves as "guardians of the Dutch landscape."[34]

Studies found that Dutch agri-environmental schemes have increased bird numbers and diversity. There have also been knock-on benefits for farmers. Larger kestrel populations are likely to help control voles. Flower strips provide food for pollinators.[35]

The impacts of agri-environmental schemes tend to be mixed. However, the Dutch approach, collectively empowering farmers as participants, stimulated a greater amount of buy-in. As with TURI, a process that was once stiflingly bureaucratic and alienating became an opportunity for trust building and developing win-wins. Could it help in other places where conservationists and farmers come into conflict?

Livestock attacks drive rural resentment of urbanites, environmentalists, and governments. Ranchers feel that wolves have been imposed on them by people who would never tolerate large carnivores in their own backyards. Remuneration for livestock losses tends to be paltry, both financially and in light of the time and emotion they invest in their animals. Nor are the psychological costs of predation properly accounted for. The bureaucratic process is usually daunting. Ranchers must submit unequivocal evidence that that their animals died from a wolf attack, even after their carcasses have been picked over by coyotes and vultures. Livestock farmers without higher education often don't even apply.[36] It is unsurprising that cattle ranchers and sheep herders see a rifle as an appealing solution to what ails them, even if some research suggests that wolf packs actually respond to lethal control by making even more pups.[37]

There are efforts underway to promote a more peaceful coexistence between carnivores and livestock farmers. One notable example is the Wood River Wolf Project in Idaho, which encourages a range of nonlethal control strategies. Better tracking of wolves can help farmers steer their herds clear of danger. Guard dogs, automatic spot lights and noise makers, attaching brightly colored flags to fencing, and other strategies can reduce predation rates and give ranchers more peace of mind.[38] They are TURI-like solutions, aiming to reduce hazards rather than forcing farmers to prove harm.

These efforts, nevertheless, are expensive and more demanding than a bullet. Many states pay for lethal control, but there is far less policy support for coexistence strategies. And wolves exist within a broader politics of rural mistrust, something that is not helped when heads of regional environmentalist groups declare that "coexistence" is a "delusion" and argue that it is livestock that need to go.[39]

Moving biodiversity back toward the zone of normal politics will require better fostering cooperation between farmers and conservationists. Biologist Henrique Pereira and I have proposed establishing TURI-like institutions to help farmers adjust their practices to support biodiversity, such as by implementing agri-environmental schemes, rewilding parts of their fields, and finding ways to better coexist with carnivores. These biodiversity extension offices would be places where government scientists and farmers would learn from one another, working together to develop solutions that aid both biodiversity and rural livelihoods.[40] But such assistance needs to be matched with the right financial incentives.

One under-explored strategy for addressing carnivore conflicts is performance payments. In Sweden, reindeer herders are paid by the government for the birth of new wolverine cubs in their districts. This more indirect way of compensating for livestock losses incentivizes farmers to better protect their herds, so that they can pocket the difference. It implicitly sets up herders to be in the business of raising wolverines as much as reindeer. The result has been a dramatic increase in the reproduction of wolverines.[41]

California has recently experimented with directly compensating livestock farmers for carnivores' presence. But it remains unclear whether western state governments will make the long-term commitment to pay tens of thousands of dollars in compensation to every eligible rancher. Even if a more collaborative form of carnivore conservation is less costly overall when compared to the current state of legalistic antagonism, today's costs are widely distributed and don't show up as a line item on a state budget.[42]

A more cooperative structure, however, isn't a panacea. In the case of Dutch agri-environmental schemes, farmers felt that the continued emphasis on monitoring and auditing reflected governmental mistrust of their efforts. Some conservationists saw the program as inherently inadequate, failing to compensate for all the biodiversity that farmers "mess up" with their crop fields.[43]

But it is wrong to judge environmental policies solely by their ecological impacts. A small change in how a regulation is organized can make all the difference when it comes to building public buy-in. The fact that some Dutch farmers now say "For me as a farmer, it is important to also take care of nature" makes the Agrarian Nature and Landscape Management policy a success story.[44] Whether it be climate change or biodiversity, we're in for a long haul. Achieving environmental goals will require starting small, reducing polarization, and building the mutual trust necessary to achieve durable political compromises.

COMMUNITY CONSERVATION

Cecil was a main attraction at Hwange Park in Zimbabwe. The thirteen-year-old lion had been studied by scientists for years. By then accustomed to people, he allowed vehicles to get close to him. He was very photogenic, described by a researcher as "powerful, but regal at the same time."[45]

Then, he was killed. American dentist Walter Palmer shot Cecil with a bow and arrow. Aided by two local guides, whom Palmer reportedly paid $50,000, Cecil was lured out of the park with bait so he could become a trophy kill. Uproar spread throughout social media and across the world. Amid death threats and protests outside his home, Palmer temporarily closed his dental practice. American late-night personality Jimmy Kimmel even choked up on camera when discussing the event. Yet, most ordinary Zimbabweans had never even heard of Cecil, and many felt that the world cared more about a dead lion than the country's endemic poverty.[46]

Although trophy hunting remains controversial, in Namibia it has founded a conservation success story. The practice funds "community conservancies" in the large, arid, and sparsely populated African nation. These organizations task locals with protecting wildlife.[47] Community conservancies are a big step away from the "fortress conservation" model that has evicted rural populations around the globe. However, as with other solutions in this chapter, they are built upon thorny compromises and face multiple challenges.

The main argument for the community conservation model is an economic one. By giving locals a stake in conservation efforts (i.e., tourism dollars) and the ability to stay on the land, stewarding wildlife ceases to conflict with rural livelihoods. Twenty percent of Namibia is now community-conserved land, bringing in $50 million to the national economy and assuring healthy populations of zebra, elephant, rhino, and lion.[48]

But community conservation isn't only about financial incentives. It is also about giving local people a say in how the land is managed. In Canada, comanaged protected areas have become the main strategy for First Nations peoples to reassert control over their traditional lands. The recently created Thaidene Nëné Indigenous Protected Area is comanaged by the national government and the Łutsël K'é Dene First Nation, allowing for a continuation of traditional hunting, fishing, and trapping—along with the promotion of tourism.[49]

Another line of argument presents Indigenous and rural populations as, by nature, conservation saviors. Agnes Kalibata, a United Nations special envoy, has argued that "indigenous peoples . . . preserve 80 per cent of the remaining biodiversity."[50] The claim is repeated by the World Wildlife Fund, the Convention on Biodiversity, and countless other organizations. Native peoples have been transformed from intruders despoiling pristine nature into environmental guardians.

This newfound status is partly a historical accident. Indigenous-controlled lands are almost by definition places that weren't colonized or lie outside the globalized economy. Some of these lands are biodiversity hot spots because of their isolation and because locals are small in number, face considerable resource constraints, and lack "modern" technology, not necessarily because they possess an inherent conservation ethic.[51] These areas have remained untouched by the forces of global economic development, forces that would typically mean either the eviction of local people or dramatic changes to their way of life.

It is also true, however, that many Indigenous groups have honed their hunting, fishing, gathering, and logging practices to be reliably sustainable. They have developed what scholars now call "traditional ecological knowledge" (TEK).

Examples of TEK are plentiful. First Nations groups in the American Pacific Northwest grew food in forest gardens of fruit and nut trees, which today remain more biodiverse than the surrounding forest. Collaboration with Inuit peoples uncovered significant deficiencies with scientists' methods for counting bowhead whale and caribou populations. Populations of both were found to be sustainable by using Indigenous knowledge about these animals' behavior. Australian governments now work with Aboriginals for bushfire management, learning how traditional burn practices better worked in tune with the life cycles of local plants and animals.[52]

Nevertheless, transforming Indigenous and rural peoples into the solution to biodiversity decline risks idealizing them. Like the blue-colored Na'vi in the movie *Avatar*, they are presented as living in a state of pure ecological harmony. Noble savage mythologies reduce Indigenous groups to symbolic foils for Western ecological corruption.

That myth also overlooks how Indigenous peoples have sometimes mistreated other species. Plains Indians often drove whole herds of buffalo over cliffs, taking the best meat and leaving the rest to rot. Some native groups thought that if any buffalo survived a hunt, they would warn other bovines to avoid humans. Others believed that buffalo emerged from underground in endless numbers.[53]

Indigenous peoples have sometimes even hunted species to extinction, as was the case for the Moa when the Māori settled in what is now New Zealand.[54] The Cree of subarctic Canada have overhunted caribou on at least two occasions: in the early twentieth century, when they acquired repeating rifles, and in the 1980s, when the species had rebounded to the point that they seemed inexhaustible. In community meetings, elders had the job of reminding hunters of their own, albeit temporarily forgotten, conservation ethic.[55] Upon obtaining axes and losing their slaves, the Yuqui of Bolivia began cutting down mature trees to obtain fruit. When asked about the practice's destructiveness they replied, "only slaves climb trees."[56] In short, although Indigenous and rural peoples have no doubt built up considerable knowledge about sustainable living in their local environments, they remain fallible human beings like the rest us.

Another problem is that the conservation ethic of many Indigenous groups doesn't exactly align with Western preservationism. During one of my classes, a Diné student pointed out that his native language doesn't refer to "wild animals." Instead, they were simply considered to be "food." Rural peoples steward rare plants because they are useful for building and decorating and as food or medicine, not for their biodiversity value per se.[57]

As William Cronon explained, wilderness is typically an alien concept to peoples who live off the land.[58] As a result of this inherent discordance, the newfound alliance between biodiversity conservationists and Indigenous peoples will continue to be an uneasy one.

That uneasiness only grows as rural peoples face choices about how to relate to the global economy and use new technologies. Indigenous peoples are diverse. Some maintain very traditional lifestyles, whereas others drive pickup trucks and use the Internet. The Saami of northern Finland continue their tradition of reindeer herding, albeit on snowmobiles and guided by GPS.[59] As eco-centric as traditional environmental knowledge can be, Indigenous peoples encounter the same dilemmas as the rest of us about how to tread on the earth.

Ignoring these complexities risks instituting a new, albeit softer, form of eco-colonialism. The Squamish of British Columbia, for instance, provoked controversy when they decided to build high-rise condos on a recently regained ten-acre plot in the middle of Vancouver. Some local suburbanites shamelessly cited the project's supposed conflict with an "Indigenous way of building" in order to prevent, well, an Indigenous group from constructing what they wanted.[60] Similarly, the Mundas, an Indigenous group in India, wanted to cut down nearby trees in order to curb elephant raids that destroy both their homes and their crops. But policies put in place to "protect" their traditional way of life prevented them from doing so. Some local activists were even pushing for laws to prohibit Mundas people from taking factory jobs. Anthropologist Alpa Shah refers to these paternalistic efforts as "eco-incarceration."[61]

William Cronon long ago warned about sentimentalizing Indigenous peoples, for, invariably, "they do something unprimitive, modern, and unnatural, and thereby fall from environmental grace."[62] The more important justification for community conservation comes from the principles of democracy and self-governance. Indigenous and local peoples ought to comanage their landscapes because they live there and care about them, full stop. To make their inclusion conditional on maintaining a proper level of traditional living, nativeness, or being in possession of special ecological knowledge is as colonial as the fortress conversation mentality that once sought their eviction. Reconciliation biodiversity means learning to simultaneously implement diverse, sometimes incongruent, ideas about what living with nature looks like.

The tensions surrounding community conservation are as visible in Namibia as anywhere else. Foreign firms have arrived unannounced to drill and mine on conservancy land, with the national government sometimes looking the other way. Some charge the long-ruling SWAPO party with complacency and corruption. Yet, in all fairness, still-developing economies can't run on conservation and eco-tourism alone. Moreover, community conservancies haven't exactly worked as planned. After subtracting guard salaries and other costs, the income to locals can be meager. Most trophy-hunting revenue is still earned on private, primarily white-owned, land.[63] In Kenya, the model has created new conflicts over land use, made it harder for farmers to deal with troublesome wildlife, and provided only modest financial benefits to locals.[64] Community conservation on its own can't reverse a country's underlying economic dysfunctions and political problems. The practice's durability depends upon healthy governments, strong democracies, and genuine economic opportunities.

REGENERATION

What was once an overgrazed dairy operation has now become a paragon of rewilding. Cattle still roam the 3,500 acres of Knepp Estate, but they're joined by wild pigs, beavers, stork, nightingales, turtle doves, hedgehogs, and countless insects. Isabella Tree and Charlie Burrell shuttered their dairy operation in 2000, letting the land go feral. As traditional farmers, they long carried far too much debt while making too little money. Inspired by the work of Frans Vera, the architect of the Oostvaardersplassen rewilding experiment, they tore down fences and stocked fewer cows.[65]

The resulting landscape is more beautiful and considerably less contentious than the Oostvaardersplassen. Tourists flock to see the rewilding exemplar. Tree and Burrell now make a much better living selling meat from animals shot on the estate, buying farmland to convert for biodiversity and carbon credits, offering safari-like tours, and running a farm shop on the property.[66] Future farmers might see supporting biodiversity and sequestering carbon to be an ever bigger part of their jobs.

What to do about agriculture remains one of the most contentious areas of the debate regarding how to live in better harmony with nature. Some people continue to insist that humans are best left out of natural landscapes. That said, proposals for sharing the land with nature, such as agroecology, agricultural rewilding, and regenerative farming, have plenty of advocates. And they arguably offer the best hope for reconciling competing environmental visions.

Knepp Estate is a case of agricultural rewilding. Landowners substitute domestic animals in an arrangement that tries to mimic a vision of a prehistoric (or at least premodern) landscape. One difficulty with these schemes is that they're mostly about rewilding. The agricultural part is more of an afterthought. They might be best thought of as ecologically conscious game reserves. According to George Monbiot's back-of-the-envelope calculations, even the entirety of the United Kingdom's farmland converted to Knepp-like rewilded land wouldn't produce enough beef to meet British demand.[67] This is where land-sharing proposals meet an inconvenient question: Can it feed the world?

Regenerative agriculture, in contrast, is more focused on farming. In fact, it is in that direction that Tree and Burrell intend to take Knepp Estate. In this arrangement, cattle are regularly moved between different pastures, allowing the grass to rebound. Farmers then bring in chickens to pick the bugs out of the cow dung and make their own fertilizing contribution to the grassland. The new Knepp will be able to stock four times as many cattle, with the biodiversity impacts lying somewhere in between that of rewilding and intensive farming.[68] I couldn't find unambiguous data regarding the potential impact of a widespread shift to regenerative agriculture to the UK's beef, poultry, and egg production, especially given that most British cows already graze their local pastures during the summer months. Critics in the United States argue that there is simply not enough land to shift to wholly grass-fed beef. However, their opponents contend that using the millions of hectares currently used for growing corn for cattle feed and ethanol biofuel would do the trick.[69]

Other proposals for more biodiverse agriculture fall under the term "agroecology."[70] These farming techniques look a lot like agriculture before the Green Revolution. They can be as simple as planting multiple crops together, such as sowing a legume species that fixes nitrogen into the soil alongside nitrogen-hungry crops.[71] In the developing world, so-called push–pull pest management is a popular agroecological technique.[72] It involves intercropping plants that naturally release pest-repelling chemicals (the "push") while seeding species that pests like to eat on the outskirts of the fields (the "pull"). Another strategy is silvopasture, where grass, trees, and animals are raised in the same area.[73] Grass and tree products feed the animals, trees provide shade and help with water retention, and animals fertilize the trees and grass. Iberian ham, for instance, comes from Spanish Dehesa silvopasture, which produces acorns and corkwood alongside the world-renowned cured pork.

Advocates of land-sparing agriculture are skeptical. They see regenerative grazing as land hungry, and agroecology as insufficiently productive and driven more by romanticism than economics. In Africa, agroecological techniques are popular mainly because Green Revolution pesticides, seeds, and fertilizers are unaffordable. One Ugandan researcher has therefore concluded that agroecology can't be more than a stepping stone to "modern" agriculture, something like what is practiced in the United States. Otherwise, it risks becoming a set of chains condemning Africans to remain subsistence farmers.[74] Critics, however, point to the costs that can come with achieving high yields with Green Revolution technologies: water pollution and scarcity, degraded soils, and indebtedness.[75]

Those arguments set up a false opposition. It has taken decades of research and development to improve the efficiencies and yields for America's vast monocrop cornfields. Something similar might be possible for regenerative grazing and agroecology. When the engineering students in my classes repeat this false binary, I ask them:

> In an era of artificial intelligence and other technological wonders, what prevents us from developing small, maneuverable, and computerized tractors that can seed and harvest intercropped fields or between rows of silvopasture? Why couldn't we genetically engineer crops that enhance push–pull pest management?

Many of the possibilities for feeding the world while also making more room for nature simply remain under-explored.

Although Knepp Estate continues to spark wonder, it remains limited as a model. Charlie Burrell is a low-level aristocrat, who inherited the land—and its associated castle—from his grandparents. Hard-scrabble farmers without such privileges will have a more challenging path ahead of them. The tourism that helps fund Knepp is due in no small part to its uniqueness. Were more ecologically friendly agriculture to become the norm, farmers could only hope to capture a small slice of the demand for landscape safaris and luxury farmland yurt rentals.

These alternative models, moreover, butt up against an agronomic system oriented toward land-sparing farming and against farm lobbies and farming cultures interested in maintaining the status quo.[76] When the problems of biodiversity and food security seem as urgent as climate change, it can be tempting to conclude that there's one "correct" solution to feeding the world. Seen from another angle, it becomes *even more important* to support a diversity of experiments, to push even harder for models that challenge and advance our thinking, for otherwise we might miss the most fruitful pastures for reconciling human needs with biodiversity. And the same reconciliation needs to happen outside farm fields, in the industrial and urban landscapes where most people live and work.

ECOSYSTEM SERVICES

At an industrial site near Seadrift, Texas, wetlands have become part of the chemical process. The Dow chemical plant produces plastics and chemicals used in antifreeze, polyester, paint, and beauty products. Part of its daily output is some five million gallons of industrial wastewater, which is processed through a set of treatment ponds. The firm struggled to meet discharge standards. Algae growth was through the roof, and hydrochloric acid often had to be added to balance the water's pH.[77]

The chemical company's own analysts found that an artificially constructed wetland was actually the best option to prepare treated wastewater to flow into Mission Bay. Not only was it cheaper to construct than a regular treatment facility, but it also required less maintenance. The wetland handled the excess nutrients and algae problem without much oversight. This came with a knock-on benefit to wildlife. Migratory birds, fish, deer, and other animals now occupy the 110-acre site.[78]

The ability of wetlands to filter water is called an "ecosystem service," a benefit that humans obtain from nature. It isn't the only one. Bees and other species pollinate our crops for us. Mangroves reduce storm surges. Plant biodiversity and herbivore grazing lessen the intensity of wildfires.[79] And then there is the simple joy that people feel just by being in natural spaces. Greater recognition of these services has driven interest in accounting for them when planning for development. For instance, Dow's interest in wetland development wasn't a one-off. In partnership with TNC scientists, they found that reforesting areas around plant sites could reduce ozone pollution.[80]

That said, it isn't exactly straightforward to then conclude that biodiversity per se has an economic value. It is true that studies have found that more diverse grasslands resist invasion, tolerate drought, and produce more biomass.[81] However, biologist Ken Thompson is skeptical, arguing that "having the *right* species is important, but having lots of species isn't."[82] Nitrogen-fixing plants matter for more for grassland productivity than rare grasses that leave conservationists in awe; resisting drought or reducing fire risk is about having the species conducive to doing so; and salt marshes filter water while being fairly nondiverse places. Even then, the relationship between biodiversity and ecosystem functioning generally displays declining returns. Most of the gains come from the first dozen or so species. Rarer, and more likely threatened, species sometimes aren't all that essential.

Nevertheless, other scientists argue that biodiverse ecosystems more reliably provide their services in the long term and in the face of drought, grazing, and other disruptions.[83] Rare species are likened to an insurance policy that kicks in during tough times.

There are, however, challenges in getting nature paid for. Assessing its value isn't all that straightforward. There's a perverse incentive to seek out ecosystem payments for lands that aren't actually under threat, which means that a firm or community effectively gets paid to continue doing nothing. Compensating locals to protect an ecosystem service, such as by refraining for cutting down trees, may lead them to dedicate themselves to different but no less damaging ways of earning a livelihood. If there isn't enough monitoring or mechanisms to ensure compliance or if payments stop, conservation may not actually happen.[84]

As journalist Emma Marris argues, the idea of ecosystem services is inherently a compromise of values.[85] As much as some conservationists try to push the square peg of biodiversity's intrinsic worth into the round hole of economic value, there are unavoidable incongruities. Not all ecosystems are economically valuable. For instance, twelve kilometers of salt marshes outside of a Dow facility didn't actually provide enough storm-surge protection to justify building smaller levees. Although ecosystem services play a role as a conservation tool, they don't make for a complete vision for nature protection.[86]

That doesn't stop some conservationists from trying to square the circle, to fit the concept of ecosystem services into the fragile nature story. Some biologists reason that, because ecosystems provide humans with different services, climate change and biodiversity decline will "cause the extinction of [humanity's] life support system."[87] When investigating such statements, I find studies that, although concerning, don't unequivocally support such catastrophic conclusions. For instance, although the diversity of pollinating insects is decreasing, generalist pollinators are thriving. Whether the resulting threat is an existential one remains unclear. Recent studies have found evidence of harms to crop yields.[88] But that doesn't necessarily mean that there's a tipping point where pollinator diversity losses drive agricultural collapse. In any case, it is enough to say that there are uncertain but potentially dramatic consequences from a world without wild species and the services they may already be providing us. We're likely better off if we don't find out what those consequences exactly are.

Yet, one shouldn't overlook that a big reason why ecosystem services strike many environmentalists as wrongheaded: The concept instrumentalizes something they consider to be sacred. And it risks subjecting nature to the whims of economics and markets. To convert species and landscapes into dollars and cents is to overlook something essential about humanity's experience of wild spaces. However valuable seeing pristine nature as sacred may be for motivating political activism, it can make it near impossible to compromise with those who don't share that vision or who disagree about how to revere the environment.

The idea of ecosystem services helps us better recognize the usefulness of protecting wetlands and other ecosystems that filter water, reduce erosion, and temper storm surges. It provides a non-apocalyptic motivation for conservation. It can help broker conflicts between people with very different worldviews regarding the environment, business, and industry. Although the value of nature can't be reduced to simple economics, the ecosystem services argument reminds us that the environment often provides us with a kind of natural infrastructure, if we let it.

GREENING THE GRAY

Bishan Park is the "Central Park" of Singapore (figure 6.2). [89] Although much smaller, only around sixty-two hectares, it provides a similar respite from the noise and heat of the surrounding concrete jungle. The Kallang River runs

Figure 6.2
Aerial view of the reconstructed Bishan Park.

through it, but prior to 2012, a person would have hardly recognized it as a stream. Like Los Angeles's eponymous waterway, it had been channelized in concrete. As part of the government's Active, Beautiful, Clean Waters Programme, the concrete was hauled away and the Kallang was restored to its traditional meandering state. Restoration included creating rock beds as well as adding reeds and other vegetation.[90]

The redesigned Bishan Park now offers several ecosystem services. During storms, the park directs the flow downstream, handling 40 percent more water than the concrete canal could. Its wetlands filter and clean rainwater. Its habitats support a biodiverse set of birds, insects, and plants. Singaporeans now repose on its grasses and walk barefoot through the pebbled streambed.[91]

Recasting nature as infrastructure expands our imagination regarding how to solve urban problems such as stormwater management. However, I think it has an even more important role for environmental politics. William Cronon recognized decades ago that the pull of "the pristine," of big wilderness, often leads us to overlook the wildness found in our own backyards.[92] We underappreciate the potential biodiversity gains to be had in and around human settlements. Sustaining wildness need not be a responsibility solely foisted upon rural dwellers but rather could become something that nearly everyone participants in.

So-called nature-based solutions, such as Bishan Park, are often seen as demonstrating nature's value to people, but they also illustrate people's value to nature. Accustomed to dismal narratives where humanity is akin to cancer or a parasite on Gaia's body, we forget that, for nonhuman species, habitat is habitat. It doesn't matter who provides it. Consider a study that, contrary to expectations, found that the intensification of agriculture wasn't the biggest driver of declines in bird populations in rural Poland. Rather, birds were being deprived of nesting sites by more modern building designs. When large, specialized farms replaced family-run homesteads, birds also lost access to small garden plots where they would forage for food.[93] Humans produce ecosystem services, and we could redesign our settlements to do an even better job of providing them.

Greening stormwater infrastructure isn't the only low-hanging fruit. Interest in so-called green roofs has grown for decades. Covering buildings in a layer of soil and plant life is great insulation, helping to lessen energy loads. Green roofs also retain rainwater, reducing the burden on sewer systems. They mitigate the urban heat island effect, which is caused by concrete retaining too much of the day's warmth. Finally, green roofs have the potential to support insect and bird populations and thereby biodiversity. Cities in the UK, Germany, and the Netherlands, for instance, have been building "bee bus stops," planting an array of wildflowers and other pollinator-friendly plants on humble bus shelters. Nevertheless, researchers warn that green roofs and other infrastructure usually aren't exactly replacements for ground-level grasslands and fields.[94] Flowers on bus shelters aren't going to bolster native rodent populations or provide homes to foxes. Still, the incremental gains are not something to play down. Small improvements multiplied a thousand- or millionfold add up to something significant.

Replacing gray infrastructure with something greener comes with its own challenges. It entails a paradigm shift in knowledge and governance. Civil engineering must be mated with knowledge of botany and ecology. Public officials need to be incentivized to make the shift. And many instances of green urban infrastructure require cooperation for government offices that were previously siloed. Nature-based stormwater management requires that environmental departments talk with transport, finance, forestry, and other agencies, and demands close work between cities and regional governments.[95] Disrupting traditional ways of building infrastructure involves a lot of discomfort.

As I showed in the last chapter, greening cities creates more opportunities for conflict between people and wildlife. Work on Bishan Park had hardly ended when a family of koi fish-eating otters took up residence.[96] Green infrastructure provides refuges for coyotes, raccoons, and other animals comfortable with city living but who are also wily and inconvenient for humans. Given that even the most urban-adapted species still try to avoid people, reducing the probability of conflict may be as simple as building more wildlife corridors. In Banff, Canada, a series of green overpasses has reduced highway collisions with animals by 80 percent. And a similar series of culverts in Queensland, Australia, has given koalas a safe means to bypass roadways.[97]

In the case of Bishan Park, locals worried that the meandering river would provide a refuge for an undesirable native population: mosquitos. Outbreaks of dengue still happen on the island. Officials reassured Singaporeans that keeping water constantly flowing and supporting populations of mosquito predators could mitigate the risks.[98] Failure to deliver, however, could undermine future trust in nature-based stormwater management. Key for ongoing acceptance of green infrastructure will be city governments' ability to be responsive to citizens' diverse needs, not only achieving goals for nature.

Focusing biodiversity efforts on cities forces urban dwellers to have more skin in the conservation game. It also exposes a broader portion of the population to the conflicts and dilemmas involved when trying to make more room for nonhuman species. My hope is that, by doing so, more urban-dwelling nature lovers will come to appreciate the value of trying to democratically reconcile seemingly incompatible environmental interests and worldviews. Biological conservation could thereby become something that works well enough for almost everyone most of the time.

NAVIGATING POST-NORMAL BIODIVERSITY POLITICS

Michael Rosenzweig's concept of reconciliation ecology continues to be cited in scientific journals and news articles. As a guiding environmental philosophy, however, it is overshadowed by calls for protected areas and by the dueling visions of economic degrowth and ecomodernism.

Like elsewhere in politics, environmental problems are often a battleground between grand ideals. That's probably unavoidable to some extent. Nevertheless, this chapter has been a plea to focus more on the mundane work of reconciling people's competing worldviews, interests, and needs. It has been a hymn to incrementalism.

Too often we believe that the scale and urgency of our problems demand radical solutions. The trouble is that incremental change is usually all we can hope for because people fundamentally disagree about what progress looks like and how to realize it. Radical breaks with the past, in any case, are usually incompatible with democracy. Big changes are typically possible only when governments ignore the cries of political losers. This is why radical upheavals tend to happen after violent revolutions or in authoritarian regimes.

Incremental change might sound trivial or inevitably insufficient. But I think that viewpoint fails to appreciate how slow and steady advancements can often outpace radical policy proposals. Big plans often either get mired in partisan gridlock or are reversed as soon as the prevailing political winds shift.

In polarized societies such as the United States, politics too often gets cast as a struggle between good and evil. "They" are like the malevolent galactic empire in *Star Wars*, whereas "we" are akin to the plucky rebels. And we can't do politics or compromise with our "enemies," whether they be climate or biodiversity skeptics, the fossil-fuel industry, or whoever. But as Phillip Wallach points out, the strength of republican institutions "is their ability to keep real, flawed, intemperate people talking, not to gradually filter evil out of our national soul."[99] Biodiversity democracy would be no different.

Win-win solutions get disparate groups to start (and keep) talking. Biodiversity offsets reduce the stakes of conservation, lessening the number of all-or-nothing conflicts between development and biodiversity. Ecosystem services attempt to partly and imperfectly reconcile competing values by showing how nature-based solutions can often replace gray infrastructure. Agro-environmental schemes and community conservancies are small, halting steps toward a more cooperative model of biodiversity conservation.

Certainly, all these approaches have their flaws. And none of them could achieve the biodiversity outcomes that would be possible if a third to one half of the globe was magically converted into protected areas. Compromise solutions will probably allow some species to go extinct. They won't protect all the habitats and landscapes that conservationists treasure. Some of nature's unique beauty may be lost forever.

The point of reconciliation, however, is to achieve not just ecological results but also social, cultural, and political ones. Only a small percentage of people are committed environmentalists, those who say they're open to making big sacrifices for nature.[100] And they won't ever have the political capital to get exactly what they want. Decades of effort to convert people into environmentalists has actually decreased the number of Americans who identify with the label. Incrementalism, win-wins, and reconciliation are necessary in order to reach people who aren't diehards. These strategies provide them with evidence that conservation can *work* for them.

Rosenzweig recognizes this. He touts reconciliation ecology as a "great environmental educator." By creating pockets of wilderness wherever possible, a greater percentage of people can come to enjoy nature's pleasures. That is, they will so long as a wilder nature isn't simply imposed upon them. "The species we live with will resensitize us to [nature's] delights and addict us to her bounty," argues Rosenzweig. "We will wonder how we ever could have settled for less."[101]

Social critic Christopher Lasch long ago pointed out the fundamental error often made by environmental doomsayers: "Too often they substitute an abstract interest in the future for the kind of palpable, emotional interest that enables people to make sacrifices on its behalf."[102] Reconciliation ecology recognizes that treading more lightly on nonhuman nature will matter to most people only insofar as they actually experience the benefits of a wilder world. And perhaps they might even come to have something like a sacred experience of nature when viewing or walking through these small pockets of human-assisted wildness.

In short, working to reconcile biodiversity with the rest of human life is a strategy for building positive constituencies for conservation. It is an act of coalition building. The point isn't to achieve consensus about ecological harmony. That is impossible. Rather, our task is to find better ways of coexisting *despite* deep-seated disagreements about nature.[103] Perhaps more transformative socioenvironmental changes will eventually be possible, but it will only be by working to bring everyone along for the ride—and letting them have a hand on the steering wheel.

Big-tent environmentalism will enhance our collective ability to propose and experiment with biodiversity win-wins. It might even open the door to more radical or novel solutions. Consider proposals to reintroduce camelids to North America, where the predecessor of modern camel species originated.[104] As mixed grazers and browsers, camelids rarely compete with other plant eaters. They are drought and temperature hardy and can eat thorny plants. By combating the encroachment of shrubs, their feeding patterns could reduce wildfire risks. Camel meat is low in saturated fat and cholesterol, and the animal's milk is rich in vitamin C. Prepared correctly or mixed with beef, its meat is a delicious source of protein.[105] No doubt that there are risks of camelids spreading invasive plants or overgrazing native vegetation. And normalizing the consumption of camelid meat and milk outside of Somali and Arab immigrants is no simple task. Nevertheless, pursuing this path holds the potential to diversify America's stock of wild and domestic herbivores, potentially restoring lost ecosystem functioning.

At the same time, desert-adapted animals such as camels, alongside kangaroos, are more efficient, providing more meat and milk for less water, carbon emissions, and feed than cows.[106] They offer an alternative to people who might be worried about the climate but are not ready to give up red meat.

More radical biodiversity proposals, such as reintroducing camels, are worth considering. They may also be more feasible than we think. Most importantly, the political work necessary to experiment with and perhaps implement them will also help us learn to do biodiversity democracy even better. That, however, will require tempering conservation's traditional conservatism.

7 RADICAL CONSERVATION

In Sakha, Russia, local men blast away at the permafrost with fire hoses, hoping to strike it rich. It's not diamonds or silver that they are looking for but rather ivory. A new gold rush for mammoth tusks is underway. It has been sparked by climatic changes that have unlocked long-extinct animals from the frozen earth, a shift that might eventually bring about these species' own resurrection.[1]

The region is isolated, sparsely populated, and poor. Twice the size of Alaska but with only a million inhabitants, monthly salaries can be as low as $500 per month. Unearthing a single tusk, however, can earn an ivory hunter more than $30,000.[2]

They aren't the only ones looking for mammoth teeth. Scientists are also there, digging in the permafrost, but not for the ivory. Their quarry is long-lost genetic information. A recent find allowed geneticists to sequence the DNA of a mammoth that lived a million years ago, which shed light on the evolutionary development of the ancient animals.[3]

For geneticist George Church, mammoth DNA is valuable less for understanding the past than for what it offers the future. He leads an effort to bring the species back from extinction. His start-up, named Colossal and cofounded with serial entrepreneur Ben Lamm, has already raised more than $400 million for the undertaking.[4]

Asked in 2019 how long it might take to resurrect the mammoth, Church responded, "Probably a decade."[5] The Harvard scientist sports a mop of white hair and similarly thick beard. Although his appearance is equal parts Albert Einstein and Prospector Pete, Church speaks in a measured and professorial baritone, evincing the seriousness with which he takes his research.

The idea of de-extinction immediately raises ethical questions. News articles on the topic usually ask, "Should we?" and reference the lessons that humanity was supposed to have learned from the Hollywood film *Jurassic Park*. Critics argue that the ethics of biological resurrection must be settled before we move forward with it. For instance, journalist Jackson Ryan insists that, prior to de-extincting any species, "we have to know if the public would even *want* that."[6]

The desirability of de-extinction, however, is just as uncertain as the science of it. Whether people can get behind resurrection biology depends on its plausibility, its eventual form, and its actual consequences. Can resurrected species return to the wild and have a good existence? Are we capable of recognizing and avoiding unintended consequences? How do our attitudes about the environment change when de-extinction becomes a readily available tool?

The motivating force behind de-extinction efforts is coming less from university researchers and more from private industry—namely, biotech start-ups. That raises concerns about how well the public's interests will be represented. The founders of Colossal insist that they will be "radically transparent." They will only patent spin-off technologies relevant to health care, not the resurrected species themselves.[7] Still, the need to turn a profit could end up running counter to ethical, ecological, and safety considerations. The risk landscape of resurrection biology is complicated. Can market-driven de-extinction earn and keep public trust?

We can't really know the answers to these questions without first trying. Neither the feasibility nor the desirability of resurrection biology is something that we can totally foresee ahead of time. Although philosophical arguments could help bring some precision and clarity to the debate, what we need most is a way to learn about de-extinction more quickly. Too often, the impacts of technological development are recognized too late and after doing considerable harm. The ecological destruction wrought by biological controls, such as the cane toad or wolfsnails, exemplifies this pattern. We need to be able to get better answers to our questions about de-extinction and other radical forms of conservation while we can still change course.

COPING WITH RISK

The waters in the shadow of Glen Canyon Dam ebb and wane throughout the day. The changing tide forms eddies that lap at the canyon's sandstone walls. Unlike the Colorado River that existed before the Bureau of Reclamation tamed it, the water flows in response to locals' electricity needs. It rises in tandem with the increasing hum of the dam's hydroelectric turbines and drops as people start heading to bed. This state of affairs has been great for humans, but native species have struggled. Many insects' eggs need to stay moist in order to successfully hatch. Decades of artificially imposed tides have resulted in a lower bug biodiversity than what is found on the Detroit River. Releases of cold, clear water from the reservoir bottom stress local fish species accustomed to a warmer, siltier environment.[8]

Tempering the dam's ecological impacts became a central concern of managers in the 1990s. They formed an advisory committee, which included the local Hualapai tribe, Western Area Power, and other stakeholders. However, no one had any idea how water releases should be altered in order to best reconcile the needs of different insects, fish, people, and other species. Scientific models couldn't tell them the perfect answer. How should they proceed?[9]

In light of decades of astounding scientific and technological progress, we too often overestimate the reach of human knowledge. We now carry tiny computers in our pockets. Artificial intelligence looks to be on the cusp of a major breakthrough. Air travel has never been safer—prior to 2025 we had gone sixteen years without a major plane crash involving a US airline. Yet, the complexities of ecosystems, society, and other large-scale systems continue to elude our collective comprehension.

Scholars have long thought about this problem. In the 1950s, political scientist Charles Lindblom argued that complex social problems could rarely be solved by quantitative modeling and analysis. No one has the time, information, or objective perspective necessary to find unambiguous answers to pressing political issues. Even then, they would still need to persuade opponents to agree to policy changes. The best we could do is "muddle through."[10]

Real life policymaking is pragmatic and hands-on and is most successful when change is incremental rather than radical. Successive trials of small adjustments in laws, rules, and regulations provide feedback on what works and what the public finds politically acceptable. Grand schemes typically either fail to pass or go awry. We saw an example of this in the last chapter, when the US Fish and Wildlife Service had to develop safe harbor programs, alongside other changes, which only imperfectly tame the unanticipated perverse incentives, and the angry opposition, created by an unexpectedly radical Endangered Species Act.

Later scholars, including David Collingridge and Edward Woodhouse, extended political incrementalism to science and technology.[11] They wondered about what sank nuclear energy. Why did the space shuttle fail to provide affordable access to space, and why did the program end in disaster?

They were equally intrigued by success stories. For instance, early genetics researchers adhered to a self-imposed moratorium on risky experiments until they learned how to better contain hazardous bacteria. One of the most surprising cases involved Danish tractor and farm equipment manufacturers, who somehow managed to build better wind turbines than engineering teams staffed by professionals from NASA and Boeing.

Through their research, Collingridge and Woodhouse found that Lindblom's lessons for legislators and bureaucrats applied equally well to science and technology. Failures tended to happen when people thought they understood reality far better than they really did. Twentieth-century nuclear engineers started building reactors at a gargantuan scale and pace before they really understood their safety challenges. Boondoggles occurred when too few people dominated decision-making and shunned criticism. The failure of the space shuttle program can be traced back to the Department of Defense demanding and getting a launch vehicle that fit their needs but at the expense of practicality.[12]

Success cases were the opposite. Early genetics researchers recognized that they were entering unexplored territory and made it a priority to learn about then-unknown risks, which is why we have had very few cases of genetically altered bacteria escaping the laboratory. Danish small-scale wind turbine builders didn't have NASA's engineering talent, so they were forced to learn by experience with incremental design changes. They shared their experiences making (and correcting) mistakes in an industry journal. This enabled them to leverage their collective intelligence into early dominance of the nascent wind energy industry.[13]

Incrementalist scholars argue that uncertainties and complexities often cannot be "solved." Rather, they are coped with. Fortunately, there are strategies that can help us more reliably *cope better*. Participants ought to scale up efforts gradually, make few changes at a time, prepare to learn from experience, foster productive disagreement, and work to eliminate conflicts of interest that prevent sensible decision-making. They call this strategic approach "intelligent trial and error."[14]

A form of intelligent trial and error is already practiced by conservationists and land managers. "Adaptive management" is an approach promoted and developed by C. S. Holling, a giant in the field of ecology. Like his contemporaries in political science, he emphasized the overwhelming complexity of reality, the limits of scientific models, the risk of surprises, and humanity's ultimate reliance on learning by doing.[15]

When twenty Puget Sound ferry terminals needed to be rebuilt in the 1990s, officials saw an opportunity to make learning part of the reconstruction process. Seagrasses are often as foundational to coastal ecosystems as regular grass is on land, so managers set the goal to carry out terminal reconstruction with no net loss of eelgrass. Each successive ferry terminal was treated as an experiment. They tried out different techniques for planting eelgrass, hoping to thwart the Dungeness crabs that liked to dig up seedlings. They discovered that design changes, mainly glass blocks in the walkway, could help eelgrass continue to grow under ferry terminals.[16]

Intelligent trial and error, however, isn't just about avoiding technical errors. Large-scale technologies and environmental management projects often falter in the face of popular opposition. As much as policy wonks, bean-counting bureaucrats, and engineers will be disappointed to hear it, the only reliable test of a policy or technological change is achieving public approval. Consider nuclear energy. Even though the technology is far safer than it was decades ago, it remains *politically unsafe* in the minds of many citizens.[17] The learning process was insufficiently intelligent. It included highly visible and potentially catastrophic mistakes: Three Mile Island, Chernobyl, and Fukushima. These errors signaled to the public that nuclear experts and regulators did not have their acts together. Despite the technology's ability to deliver low-carbon energy, low levels of public trust arrested the development of nuclear reactors for decades.[18]

The Glen Canyon Dam Adaptive Management Program has turned an almost sixty-year-old dam into an ongoing science experiment. Sometimes, managers release large flows of water to try to build sandbars. At other times, they keep flows steady for days at a time, hoping to spur river bugs to reproduce. Then, they watch. Scientists have reported finding more egg casings on riverbanks. Fly fishermen have complained of buggier conditions. Dam managers from around the world have come to Glen Canyon to try to learn how they too might balance ecosystem needs with demand for electricity and irrigation water.[19]

Coping with the complex politics of the Glen Canyon Dam, however, remains a work in progress. An assessment of the first fifteen years of adaptive management found that a great deal had been learned about how to reconcile water releases with the local ecosystem, but daily operation of the dam had remained mostly unchanged. Different stakeholders had become entrenched in their positions. Each side would hire their own experts to prove who was "right." A coalition of representatives from utilities, state governments, and sometimes tribes often thwarted motions to implement water releases for the environment, leading environmental groups to turn to litigation. The assessment illustrates a key consideration. All the trial-and-error learning in the world can't amount to much if the surrounding political system is dysfunctional.[20]

How could intelligent trial and error help us navigate more radical biodiversity conservation questions? When should we help endangered species migrate to greener pastures? Could we responsibly use genetic engineering to ensure their survival or to eradicate their enemies? How might we best learn the benefits and risks of bringing species back from the dead?

DAMMED IF YOU DON'T

No one knows how beavers got to the River Tay in Scotland. Some people believe that they might have escaped from an animal lover's private collection. I like to imagine that a crack team of amateur rewilders did it. Slinking through farmers' fields under the cover of darkness, they delivered the contraband rodents to the riverbank. It was a tearful farewell as the doors to the animals' crates were unlatched. But the musky mammals hesitated. "Go on then," someone whispered in a lilting Scottish accent. The newly freed beavers briefly looked back over their shoulders, flashing orange-hued toothy grins, before diving into the water.

Like their North American cousins, Eurasian beavers gnaw down trees and dam up rivers, engineering the environment to produce wetlands. Humans can gain from this activity. Beaver dams help retain water for dry times and hold back floods, although not as predictably as the man-made version. Although damming streams in the foothills isn't so problematic, low-land farmers are reasonable to worry about losing large swathes of crops to beaver-induced flooding.[21]

Beavers weren't only reintroduced to Tayside. However, it is where the animal stoked the most controversy. There was an accidental release on the River Beauly, but the animals were quickly relocated. A small population at Knapdale was explicitly reintroduced, preceded by months of public engagement and scientific study. In Tayside, a big farming area, the government decided to accept the unplanned presence of beavers on the river. Farmers were aghast, arguing that environmental laws were being applied inconsistently and to their detriment. Some told interviewers that they'd shoot the beavers themselves, given the "conspiracy" against them.[22]

The contrasting viewpoints of locals at the three reintroduction sites are illuminating. The same intervention can elicit very different responses, depending on how it is carried out. Landowners in Tayside were forced to react to the sudden appearance of beavers and felt that they lacked the government's support. Unsurprisingly, they've relied on lethal control. In order to prevent crop losses, eighty-seven beavers were culled in 2020.[23]

Things could have been different. The case highlights the need to include the people impacted by a reintroduction early in the process. Beavers could have been restored to Tayside without evoking such extreme animosity had farmers' concerns been taken in consideration.

Not-in-my-backyard (NIMBY) opposition isn't inevitable. Citizens don't rage against the introduction of high-rises, homeless shelters, wind turbines, or beavers because they are intrinsically opposed to change. Well, at least, not all of them do. Often, it is more that they resent feeling as if their lives are being controlled by others. They believe that consequential decisions were made behind closed doors or were unfair.[24]

Intelligent trial and error can help avert popular resentment. Beaver reintroduction in Tayside could have been more gradual and controlled. Steps could have been immediately taken to protect against beaver impacts. Trees could have been saved from rodent teeth by covering the base with wire netting. Fencing could have prevented beavers from spreading uncontrollably or damming up culverts.[25] Advocates could have sat down with skeptical farmers to negotiate and hammer out compromises for the species' continued existence.

Good public engagement is a process of creating a positive constituency for change while preventing the inadvertent creation of a large negative one. In other words, make as many friends, and as few enemies, as possible.

Tayside isn't a particularly extreme case. Beavers used to romp around Scotland. As far as conservation interventions go, it wasn't particularly high risk either. Nonetheless, it provides guidance for how to approach a more radical proposal: relocating endangered species across the globe.

ENVIRONMENTAL REFUGEES

When Connie Barlow was hiking in Torreya State Park in 2004, the pine tree that gave the park its name was conspicuously inconspicuous. She eventually found short and spindly Torreya pines, cut down by a fungal blight, huddled together in the park's ravines. Something about these evergreens, which were more reminiscent of Charlie Brown's Christmas tree than not, stirred Barrow's emotions. As she lay down under a sapling, she promised it, "I'm going to find a way. I'm going to move you."[26]

Barlow is a lot like those people whose kindheartedness leads them to endlessly rescue stray cats and dogs, even as their homes fill with kennels and the smell of animal waste. Except, she's gaga for trees. Barrow admitted to science journalist Zach St. George that she once swallowed alligator juniper seeds from Albuquerque's Sandia Mountains and then drove hundreds of miles north, so that she could use her own digestive system to aid in the species' "dispersal."[27] After her experience in Florida, she started a group called the Torreya Guardians, which has dedicated itself to collecting the tree's seeds and mailing them across the country.

In the face of migration pressures, such as climate change and habitat loss, many species will struggle to overcome the roadblocks that we've put in their path: miles of farmland, subdivisions, and interstates. How might humans most responsibly give these species a helping hand?

For the most straightforward cases, we might only need to help a species expand their homeland. The awe-inspiring, but fickle, California redwood is one example. The giants prefer a certain climate. Yet, it still isn't clear how temperature and precipitation on the California coast will change in the next decades. Redwood National Park, however, has set boundaries. What if redwoods eventually no longer survive within the park's borders? What if their future habitat lies in areas of Northern California and Oregon currently occupied by farms and small towns?[28]

Scientists are hard at work developing models that could provide ever more fine-grain and accurate prediction models for California's climatic future. But those models are not really key to solving the problem. Forests expand and contract incrementally. Officials could prioritize small parcels of the most promising borderlands, working to turn small strips into conservation easements and repeating the process over the decades as the climate gradually shifts. Doing so would still require difficult negotiations with landowners. To facilitate the process, officials should propose only temporary easements at first, seeking to make them permanent only once it's clear that redwoods are indeed heading that way. Moreover, landowners would need to be more enticed than forced, perhaps by offering them the ability to trade for lots from areas where redwoods are retreating.

Other cases, such as airlifting a species to whole other regions or continents, are much more complicated. Still, there's a reasonable argument for getting over our initial squeamishness. Climate changes will leave species marooned. According to some scientists, the Torreya pine should have moved north after the last Ice Age because it is far better suited to colder climates.[29] Likewise, Ethiopian wolves will eventually run out of mountaintop habitat as the world warms, unless they can be moved to cooler peaks.[30]

Migration can offer a lifeline to species whose living space is shrinking. California redwoods transplanted to New Zealand's Whakarewarewa Forest a hundred years ago are thriving. A future for that iconic species, and others, might be best assured by assisted migration.[31] But doing so raises the complexity and uncertainty. It increases the risk.

Scientists are unsurprisingly divided. Daniel Simberloff and his colleagues consider assisted migration to be "ecological gambling" and decry the risk of migrating species becoming invasive.[32] Other scientists highlight the risks of inaction. Many landscapes are already in a transition to new ecological realities. Introducing species can help ecosystems adapt to future environmental conditions. For instance, foresters have helped preserve black ash wetland forests, which are threatened by an invasive pest, by introducing new tree species, such as American elm and Manchurian ash.[33]

As much as the question of whether it is morally right to pursue assisted migration may excite philosophers, I don't think it makes for a productive conversation. Critics, such as Daniel Simberloff, point out that the risks are extremely difficult to predict. And the same is true for the potential benefits. Recognizing the staggering uncertainties shouldn't be a recipe for inaction. Obstinately insisting on staying put is no more intelligent than imprudently plunging ahead, especially when the world around you is already changing. Rather, the question is how we ought to design trials of assisted migration in order to learn if, when, and where it can be intelligently done.

A research team led by Australian biodiversity scientist Nathalie Butt warns that endangered species don't have time to wait for complete scientific understanding. Cognizant of the uncertainties, they recommend several commonsensical strategies to reduce the risks. One is to focus first on species that have an inherently low risk of invasion. For instance, trees such as red-fruited ebony take a long time to reproduce and thus spread slowly. They aren't likely to get out of control. Other species could be sterilized or prevented from going to seed so long as they remain an experimental population. Both approaches lessen the risk of invasion. Uncertainties can also be reduced by starting with shorter-distance migrations and smaller-scale introductions. For example, the Mount Claro rock wallaby only needs a helping hand to hop over miles of farmland in order to resettle in suitable high-elevation habitats. In any case, experiments would need to be intensively monitored, with well-resourced managers empowered to act quickly against newcomers that seem too "successful." The exact set of strategies used would be species dependent. In any case, the point is that gradual assisted migration could be done with sufficient intelligent trial and error to drastically reduce the likelihood and consequences of ecological mistakes.[34]

The Torreya pine is actually a pretty good candidate in this regard. It's not likely to be invasive, although some people worry that translocation could bring the pine's fungal blight along with it. Most of the opposition to assisting its migration really comes from biologists who fear that it sets a dangerous precedent.[35] To them, assisted migration threatens to undo progress in getting people to think twice before introducing species to new continents.

A widespread bias toward restraint will end up hindering intelligent trial and error for assisted migration. Most jurisdictions don't have explicit guiding rules for the practice. Organizations such as US Fish and Wildlife Service are afraid to get involved in an era when agency decisions tend to bring an avalanche of lawsuits from ENGOs. And current laws tend to maintain a distinction between native and alien species that doesn't acknowledge climatic changes.[36] Where officials are reluctant to act, informal efforts such as the Torreya Guardians will fill the vacuum. However, amateur-led assisted migrations tend to proceed without the appropriate guidance of biologists, and they aren't prepared to include and compensate people who are potentially harmed. Mistakes, much like the illicit release of beavers in Tayside, will eventually evoke local resentment.

The point of public involvement, however, isn't just to keep conflicts from blowing up. It can help push us out of a taken-for-granted status quo.

As global environmental changes put the ecosystems of yesterday under pressure, we face challenging questions about how to maintain the resilience and vibrancy of our local environments. Scientists can't answer them for us. Not only are the potential benefits of relocating species to more fruitful pastures unknown but so are the risks. Whether it is better to try to preserve today's landscapes or cultivate novel ones is unavoidably a value judgment. All science can do is help us better explore those pathways.

It is hard to say how successful Barlow's effort has been. According to the Torreya Guardians website, established groves in North Carolina seem to be doing very well. Even so, these efforts haven't been scientifically studied. Nevertheless, assisted migration has started to move into the mainstream. Managers of Minnesota's Chippewa National Forest have been experimenting with "climate winner" species from the south, such as white oak and Ponderosa pine. In this case, they are trying to save not an endangered species but rather the forest itself. As forest service scientist Brian Palik explains, "The climate of southern Minnesota has been in northern Minnesota for 20 years."[37]

The ecological baselines of the past are increasingly out of reach. The future of conservation may demand moving species around in order to guide ecosystems through human-driven environmental changes. Some scientists, however, think that even assisted migration doesn't go far enough. They believe that better adapting species to a changing world requires altering their genetic code.

ENGINEERED FOR SURVIVAL

Breeding coral requires a lot of patience. The marine animal reproduces during rare mass spawning events. These are spectacular eruptions of eggs and sperm, involving hundreds of coral joining together in a collective moment of maritime coitus. The underwater blizzard of white, red, and yellow gametes is awe-inspiring, even if it is also kind of gross (figure 7.1).[38] These events sometimes only occur once per year, when seawater temperatures, the phases of the moon, and other factors align.[39] In other words, the vibes have to be just right.

Figure 7.1
A coral spawning.

When scientists working for the National Sea Simulator in Australia succeed in provoking a mass spawning event, the laboratories become a flurry of activity. The flow of white-coated scientists comes to resemble the raging storm of coral semen and eggs. Research assistants race around the darkened rooms wearing red headlamps—they wouldn't want to kill the mood and interrupt the invertebrates in their moment of aquatic ecstasy. Pipettes are used to collect eggs and sperm, which are separated and prepared for interbreeding. It's all part of an effort to develop supercharged coral for a warming climate.[40] Scientists call it assisted evolution or facilitated adaptation.

Coral isn't a single organism. The brightly colored underwater forests showcased in nature documentaries are actually colonies of sac-like polyps attached to skeletons of calcium carbonate. And these polyps rely on a symbiotic relationship with dozens of algae and bacteria species in order to eat and live. When coral is "bleached" in response to a heat wave, we are witnessing the coral polyps suddenly treating their resident algae as if it were an intruder. Normally the algae make food for the coral, which is otherwise scarce in the nutrient-poor waters where reefs form, but the rising temperatures drive them to produce damaging free radicals. The coral eject the algae, and their color along with them.[41]

Coral's biological complexity, however, has its advantages. It gives scientists multiple pathways to try to help coral adapt to climate change.[42] One approach is to selectively bred the coral themselves. Some species have already become more resilient to warming waters. A 2017 heat wave near the South Pacific island of Palau resulted in few bleaching events, despite being more severe than a highly damaging one in 1998.[43] Coral have the capacity to evolve. Even so, climate change–driven temperature changes could be too rapid for evolution to take place, so scientists crossbred coral with others from warmer environments. Or they heat stress individuals in the laboratory, hoping that they'll pass on temperature tolerance to their offspring. A similar set of techniques is used to facilitate the adaptation of the symbiotic bacteria and algae.

The most controversial proposal, however, is genetic engineering. Isolating specific genes for tolerance to heat can accelerate the process of producing climate-ready coral. The technique's relative novelty, and the fact that it doesn't have an exact analogue in nature, increases worries about unintended consequences.

Engineering coral for slightly improved climatic resilience is unlikely to produce an ecological monster. Most cases of assisted evolution are nothing like the release of the cane toad into an unsuspecting Australian landscape. They're an already-native species that is being minutely modified.[44] Perhaps a genetically engineered version could outcompete naturally occurring coral. However, that risk could be mitigated by engineering a diversity of coral species, ensuring that not just one of them gets a helping hand from humanity. In any case, we could limit the seeding of engineered coral to smaller, more isolated reefs, so that scientists could observe the longer-term impacts.

When it comes to using genetic engineering for assisted evolution, getting the politics right is essential. That, no doubt, includes on-the-ground precautions to avoid large-scale mistakes. Scientists and environmental bureaucrats would also be wise to invite the participation of nature lovers, locals, and skeptics. Otherwise, it will be difficult to convince the public that they aren't Dr. Frankensteins thoughtlessly toying with the fabric of life.

Fortunately, we have models to follow. Consider the American chestnut tree. The species was all but extirpated by a fungus that stowed away on Asian imports a century ago. The blight causes rapidly growing cankers to form on the tree's stems and branches, causing them to die back to a stump. The tree's roots continue to sprout, only to be reinfected and knocked back to the ground.[45] As a result, a tree that once covered the eastern United States and served as an abundant source of calories for humans and animals alike is now almost completely absent.

One possible solution, recently developed by researchers at the State University of New York at Syracuse, involves inserting a gene from wheat into the tree's genetics. This theoretically enables the chestnut to break down the toxin produced by the blight. Results have been mixed. Nevertheless, the effort exemplifies ongoing learning about how to responsibly experiment with novel organisms.[46] The tree was subjected to a battery of tests. Scientists looked at whether its decaying leaves or pollen could impact soil fungi, tadpoles, or bees. Moreover, they took reasonable precautions. Experimental trees had their flowers sealed off and were restricted to greenhouses and fenced plots.[47]

To some environmentalists, the project is guilty by its association with the biotech industry. Lead scientist Bill Powell laments being accused of being "a Monsanto shill."[48] Firms such as Monsanto are broadly mistrusted. Citizens see them as getting to unilaterally decide what our food system looks like.[49] They worry that biotech companies aren't following proper testing procedures. And for good reason. A recent legal settlement against Monsanto found that the company "failed to conduct adequate studies" on the safety of its Roundup pesticide, "distorting the scientific inquiry" and being "[unwilling] to inform the public."[50]

Public mistrust of genetic engineering stems from perceived unfairness more so than its demonstrable risks. To opponents, genetic technoscience props up an industrial farming system that benefits corporate firms at the expense of small farms, biotechnologists are unjustly allowed to patent life itself, and genetically engineered agriculture may end up pushing some farmers further into debt.[51] The technology's image is inseparable from the politics of big biotech.

Scientists leading the American Chestnut Project have explicitly worked to distance themselves from the biotechnology industry. They declined to patent the gene that may save the tree and made their research publicly available online. So far, the project has yet to work with for-profit tree nurseries, preferring to remain a volunteer effort and avoid the appearance of commercial motives. Involved social scientists have even collaborated with Indigenous groups and scholars, developing words for the genetically engineered tree in Haudenosaunee.[52]

The perception that industry ties inherently corrupt biotechnology, however, is ultimately self-stultifying. Commercial firms possess the expertise and capacity to produce blight-resistant chestnut trees and heat-tolerant coral at the scales necessary to actually make a significant impact. Excluding them would virtually guarantee that these efforts never amount to more than interesting experiments.

Trust in genetic engineering, unfortunately, can't be regained by the American Chestnut Project alone. Moving the needle will require that much more serious public engagement efforts become the norm within the biotech industry. Citizens need to have an opportunity to shape consequential decisions about when, where, and how synthetic biology gets deployed. The novel technologies need to be scaled up slowly and intensely monitored by trustworthy third parties.[53]

So far, reactions to engineered coral have been mixed. When scientists sought to release a coral that had been merely interbred to resist bleaching, Hawaiian officials discouraged them from seeking a release permit. On the other hand, Australians, perhaps motivated by the Great Barrier Reef's status as a national symbol, have been more receptive.[54] It is unclear whether future efforts will get framed as a commercial biotech money grab or an honest conservation effort. Not only are coral reefs a public good but they also support commercial fisheries and tourism.[55] Commercial interests are largely unavoidable. Still, building trust in conservation biotechnology will require ensuring that those interests don't appear to dominate the process.

GENE DRIVEN TO DEATH

The mice are everywhere. One online video shows them pouring by the thousands of out a silo and into a large bucket. The ground looks like a bear skin rug that has come alive. Farmers have discovered mice making homes in every nook and cranny of their tractor or raining down in a hailstorm of little black bodies from their machinery. Every few decades in Australia, imported house mice take on plague proportions and cause untold amounts of crop damage.[56]

Across the Tasman Sea, the house mice's rat cousins are despised for different reasons. They either eat the eggs of cherished flightless birds or serve as a food source for the stoats and feral cats that prey on the birds themselves. Some of the local Māori cherish the kiore rat as spiritually important, despite its status as an invasive invader. But at the national level, the animal is conservation enemy number one. In both countries, scientists are developing genetic engineering technologies that might eradicate the rodents once and for all.[57]

The politics are clearly more fraught for this case. This isn't about bringing a species back from the brink. A gene drive is being developed in order to cleanse a class of animal from the landscape. It raises a legitimate question: Should humans be so uncompromising in deciding where species get to exist? As we saw earlier, some consider nativity to be a romantic and xenophobic idea, one that condemns successful species in order to protect those that no longer fit into today's world. In any case, gene drives are touted as a more targeted and humane way to extinguish ecological trespassers from an ecosystem. Unlike dropping poison from planes, there's less risk of indiscriminately killing other species.

Gene drives are made possible by CRISPR technology, which, in turn, enables a more precise removal or addition of genes into an organism's DNA. What makes the technique appealing for conservation is that it mostly overcomes the inconstancy of genetic inheritance. Normally, only a portion of offspring will inherit any particular gene from their parent. If you engineered a mouse to produce sterile female babies and then released it into the outback, the inheritance lottery would limit its impact on the overall population. Scientists have gotten over this hurdle by including a gene-drive sequence, which forces the organism to pass on a sterility gene whenever it reproduces (figure 7.2).[58] The result is literally a "selfish" gene, one that ensures its own propagation, even at the expense of the species.[59]

Simulated releases have found that the introduction of just 256 edited mice could eliminate a population of two hundred thousand within twenty-five years.[60] A gene drive's relentlessness, however, is a double-edged sword. By almost guaranteeing inheritance of a trait, there's a high risk that it becomes invasive. For this reason, one of the pioneers of gene-drive technology, Kevin Esvelt, is also one of the loudest voices pleading for precaution. As one magazine feature writer observed, "Not since Robert Oppenheimer has a scientist worked so hard against the proliferation of his own creation."[61] He and his collaborators worry about the consequences of a gene-driven rat escaping or being smuggled out of New Zealand. What happens to the world's wild rat populations?[62]

The risks are purely theoretical at this point, but Esvelt and other scientists are working to develop strategies for reducing them. The main approach is to encode the gene drive so that its spread is inherently limited. In a "daisy chain," parts of the gene-drive genetic code are separated and arranged in a chain, and each inherited trait triggers the next one, similar to the stages of

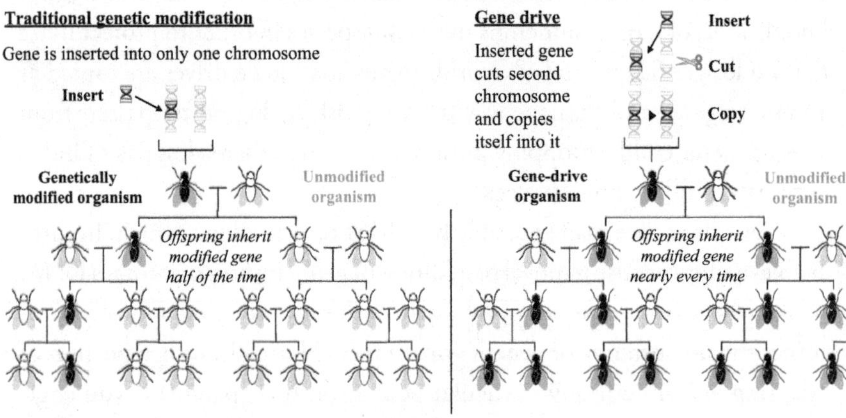

Figure 7.2
Illustration of how gene drives push genetic modifications into a population.

a rocket. Eventually the distinct components of the gene drive fall out of the gene pool one by one. In a sense, the genetic rocket "runs out of fuel."[63] A daisy-chained gene drive should be self-limiting. Nevertheless, "molecular containment" suffers from an internal conflict. The gene drive needs to spread well enough to eradicate the offending organism but not spread so well as to threaten neighboring species or innocent populations.[64]

This contradiction makes the technology inherently difficult to manage. One must walk a fine line. Governments and scientists have to carefully design and monitor experimental introductions, hopefully well enough to avoid surprises later. The World Health Organization lays out one potential multistage process for gene-driven organisms. The process starts with laboratory studies and only moves to field releases after "contained trials," where artificial walls or ecological boundaries prevent unintentional spread.[65]

As mentioned earlier, intelligent trial and error involves more than simply getting a better handle on the technical risks. There also needs to be a process for persuading the public that the effort is trustworthy. Do scientists and regulators have their stuff together? Are decisions transparent and responsive to public concerns?

Any gene-drive conservation effort will have to do better than Oxitec. The firm has pioneered engineering male mosquitos to produce nonviable female offspring. Only the females of the species bite and spread diseases such as dengue and malaria. After a relatively uneventful trial in Brazil, the firm ran into significant resistance in the Florida Keys. Although Oxitec was invited by the local mosquito control district, residents still felt blindsided. They resented being signed onto a scientific experiment without their consent. The firm's scientists insisted that the risks to the environment, or from mosquito bites, were low. Such assurances, however, fell on deaf ears. Public engagement came far too late and seemed too one-sided. Locals mistrusted safety answers coming from a for-profit company. And it probably didn't help that the trial was hosted in the unincorporated community of Key Haven, which looked like a cynical strategy to sidestep getting locals' permission. It would be nearly a decade before the EPA approved the plan.[66]

In light of these problems, advocates of gene-driven conservation should take a different tack. They might take a cue from nuclear power. Most atomic power plants are drawn from a narrow range of designs (i.e., the light water reactor). At large scales, these reactors are inherently troublesome. They need to be actively cooled. And even after they are shut down, meltdowns can occur if reactor core cooling is disrupted. Two of the world's major nuclear accidents involved major loss of coolant incidents. In Three Mile Island, a pressure relief valve was stuck open, unbeknownst to operators. Coolant stopped flowing to Fukushima's reactors because backup generators were inundated by a tsunami triggered during the Tohoku earthquake. Modern reactor designs try to get around this problem by including emergency cooling systems that should work even in the absence of electricity. Even then, citizens are reasonable to wonder "What if?" Backup systems have components that can fail if not regularly inspected or maintained. Poor management can undermine even the best laid plans.[67]

It is easy to forget that nuclear power plants need not be so hazardous. There are alternative designs that can adequately cool themselves during an emergency event. Simply shrinking the scale of standard reactors can do the trick. Such designs are *inherently safer*. They don't require an array of emergency backup systems that must work perfectly whenever unanticipated errors occur. They are more forgiving to the people in charge of assuring their safety. An early advocate of less hazardous technological design described inherent safety simply: "What you don't have, can't leak."[68] Removing unnecessary hazards is often the best way to avert catastrophe.

Gene drives could be inherently safer. If the genetic changes aren't lethal for a species, they can't potentially eradicate an unsuspecting nontarget populations. Scientists investigating ways to reengineer the cane toad, for instance, haven't limited themselves to sterilizing genetic changes. They have also looked at reducing the species' toxicity.[69] A detoxified cane toad would accomplish many of the same goals as extermination while lessening the potential cost of error. Should the modified toad escape Australia, it would probably be an easy target for predators already accustomed to the poisonous version. In a similar vein, anthrozoologist John Bradshaw has proposed sparing songbirds by engineering domestic cats to have weaker hunting instincts.[70] It could be possible, although likely more complicated, to engineer mosquitos to simply not spread malaria to humans. Enfeebling, rather than eradicating, gene drives are less risky. They are intrinsically safer because they don't require us to rely on genetic backup systems, hypothesized biological or geographic barriers to spread, or uncertain export quarantine measures to prevent the most catastrophic unintended consequences.

Focusing on enfeebling would help avoid the other unanticipated consequences of eradication. Even though mice and rats are widely maligned, they still fill ecological niches. Rodents spread seeds and feed raptors, foxes, and other carnivores.[71] What happens to ecosystems when they suddenly vacate those niches?

Finally, the inherently safer approach reflects the reality that most species usually aren't powerless victims. They're typically not like silent film depictions of helpless damsels in distress being tied to railroad tracks by mustachioed villains, waiting for a conservation hero to come and rescue them. What they need from humans may be not the eradication of invaders but rather simply more time to adapt.

Some scientists maintain that the risks of driving sterility in mouse populations are low, that the spread of genes to nontarget species is unlikely. House mice don't appear to readily breed with native mice in Australia.[72] Perhaps gene-driven sterilization is unproblematic in Oceania. Maybe rats and mice do too much harm and offer too few ecological benefits. Nevertheless, exploring alternatives to eradication is still worthwhile. Could house mice and rats be gene driven to be less cautious around birds of prey, worse at smelling out crops or flightless bird eggs, slower to reproduce, or otherwise less well adapted to the ecosystem?

Many of the conflicts over technological innovation stem from a lack of imagination. In the midst of overwhelming uncertainty, we gravitate toward simplistic stories. Sometimes, these are tales of doom or irrepressible optimism. Artificial intelligence will put us all out of work. A driverless car future is "just around the corner," same as it was ten years ago. The idea that synthetic conservation must entail eradication is a similar kind of story. By deploying an intelligent trial and error learning process, we could arrive at a technology that is both reasonably effective and exists in a form that most people, not just scientists and technologists, are comfortable with. The process transforms public hesitancy from a barrier to technological change into a resource. Citizens' resistance can spark more creative thinking.

Gene-drive pioneer Kevin Esvelt continues to work on genetically engineered organisms. He has even developed a mouse that is resistant to Lyme disease, a sometimes serious tick-borne infection affecting some half a million Americans every year.[73] White-footed mice are a major reservoir for the bacteria that causes Lyme disease. Eliminating them from the picture could help prevent outbreaks.

Locals are hesitant, but, fortunately, Esvelt is the ideal scientist for the job. Reflecting his great respect for the uncertainties of genetic engineering, his group conducts honest and humble public engagement. In response to local resistance, they have thus far foregone adding a gene drive to their engineered mouse, sacrificing technical sweetness for public acceptance.[74] They are planning an experimental release on small islands near Nantucket, Massachusetts, hoping to prove the technology's effectiveness and safety well enough to justify larger-scale tests.[75]

It isn't yet clear if gene drives will be used to eradicate alien invasive species. There is already plenty of hesitancy surrounding non-gene-drive engineered organisms. If gene-driven eradication happens anywhere, it'll be on small, isolated islands. The harm of rats, mice, and other newcomers is more clear-cut, and locals are more accustomed to extreme measures of biological control. Whether the technique gets applied elsewhere will depend on advocates' ability to prove themselves trustworthy. Indeed, what biodiversity conservation may need most in the twenty-first century are institutions to help them do exactly that.

RESURRECTION BIOLOGY

Before the thylacine was eradicated, someone preserved thirteen babies of the marsupial species in alcohol. With wrinkled beige skin and sitting in a milky liquid, the joeys look every one of their hundred years of age. Thylacines, often referred to as Tasmanian tigers, have been extinct since 1936 (figure 7.3).[76] The traditional "pickling" process for storing animal samples has been as hard on the joeys' DNA as it has on their skin. Only one of the

Figure 7.3
A thylacine photographed at Beaumaris Zoo in 1936.

thirteen jars yielded enough genetic information for scientists to map the thylacine genome.[77]

The scientists studying these samples were interested in how a species so distant from dogs could have evolved nearly all of the same traits.[78] Apart from a series of black stripes on its back, and of course belly pouches, the thylacine resembles other wild canines. It is hard to not notice the puppy-like appearance of the jarred babies (or feel sad about it). The DNA sample, however, not only helped to answer questions about the species' evolutionary development but also offered the hope that the thylacine could be revived, *Jurassic Park* style.

The effort is headed by Andrew Pask's lab at the University of Melbourne. The geneticist has also teamed up with Colossal, the start-up founded by Ben Lamm and woolly mammoth devotee George Church.[79]

Marsupials are attractive candidates for de-extinction. Offspring only spend a matter of weeks in the womb. They are no bigger than a jelly bean when they make the long trek to their mother's pouch, where the joey will suckle until it is ready to explore the outside world. The briefness of pregnancy makes it easier to use very different animals, such as the rat-like dunnart, as surrogate mothers for cloned Tasmanian tigers. Moreover, it means that the replication process can be iterated, and ideally perfected, relatively rapidly. For a mammoth, a successful pregnancy in an elephant surrogate would drag on for almost two years.[80]

The question of "Should we?" naturally arises. Even though the thylacine was ultimately eradicated by humans, they had their own problems. Some scientists think that the species was in terminal decline long before people put bounties on them and imported dingoes started encroaching on their territory. Low genetic diversity left them susceptible to disease.[81] Even if thylacine embryos can successfully gestate in a long-tailed dunnart and be reared to adulthood, it isn't clear if scientists can squeeze out enough genetic diversity to produce a viable wild population.[82] That said, help may be on the horizon. New techniques can recover more genetic information from the sliced-up DNA found in pickled samples.[83]

Pask and Lamm say that they targeted the thylacine because it was the only large marsupial predator in existence. They compare returning the tiger to Tasmania to wolf reintroduction in Yellowstone, imagining the future reintroduction of a keystone species to the Australian continent.[84] Even if that goal ends up being impractical, techniques pioneered with the thylacine could be used to help other animals.[85] Indeed, woolly mammoth de-extinction efforts have already spun off discoveries that may help protect captive Asian elephants from a deadly herpes virus. Thylacine research, for instance, might be a lifeline for koalas.[86]

From another perspective, these reasons seem more like rationalizations. I don't think it's a careful weighing of risk and benefit that really drives interest in de-extinction. Colossal's website justifies researching the topic by referencing moral decency and the firm's dedication to a rewilded planet. De-extinction is both a guilt-driven drive to "[right] an anthropogenically induced wrong" and a vision of a restored, albeit human-designed, wilderness.[87] It is de-extinction's story of redemption that captivates us.

Even so, and despite its radicality, resurrection biology isn't spectacularly risky, especially given that the focus is on large animals. It will be a major feat to produce a viable population of replicas. It was a decades-long slog to reestablish a wild population of Arabian oryx, and biologists at least had nine living animals left to start with.[88] Invasion, in this context, seems highly unlikely.

The main criticism of de-extinction holds that it is as a conservation "sideshow." The major drivers of extinction, in this view, are climate change and habitat destruction. De-extinction takes vital resources from these more important conservation efforts. As philosopher Ronald Sandler puts it, although "it is terrible that there are no longer enormous migrating flocks of passenger pigeons in the United States or freshwater dolphin pods in China . . . what is even more terrible is that this is no longer a world for them." In short, bringing back species misses the real problem.[89]

That line of thinking, however, is overly simplistic. De-extinction start-up Colossal and Stewart Brand's Revive & Restore nonprofit are funded by tech industry money. They aren't clearly parasitic on existing conservation efforts.[90] Moreover, critics underestimate the power of the story of resurrection biology, especially among people who aren't traditional environmentalists. The endeavor speaks to those who seek a positive, innovation-driven answer to the biodiversity challenge, people left out by the dismal calls for limits and demands for expanding protected areas. Criticism also overlooks that de-extinction might help build a large positive constituency for conservation. If and when we have viable thylacines or mammoths, supporters will want to find or create places to put them.

The debate over de-extinction exposes a major gap in the governance of conservation. The current mishmash of private and public efforts, all with their own disparate goals and visions of nature, does reasonably well to represent the interests of a great deal of people. On the other hand, biodiversity conservation is far too important to be steered by the chaotic whims of Silicon Valley investment, ENGO lawsuits, or environmental bureaucracies designed back in the 1970s. We need higher-level policymaking to set at least a few public priorities for conservation and to work to better reconcile enduring conflicts.

The United States, and most other countries, have no established venue for evaluating and steering efforts such as de-extinction, nothing like the Glen Canyon Dam Adaptive Management Program but for biodiversity. We should.

For such a large and contentious issue, some combination of experts, stakeholders, and ordinary people should be regularly meeting in heated deliberation.[91] It wouldn't be within this body's purview to try to answer the "Should we?" question for radical conservation efforts—it wouldn't be able to anyway. Nor would its purpose be to "kill" or prohibit certain efforts. Rather, the point would be to convert disagreements regarding biodiversity into guidance for ongoing experimentation. The more precautionary-minded would point out potential oversights or blind spots. They would agitate for more safeguards and monitoring systems for novel strategies, such as de-extinction. The more proactionary, in contrast, would try to steer governmental R&D into promising, but uncertain, conservation experiments.

The goal of such meetings would not be consensus. The quest for an unequivocally "correct" answer for any biodiversity question is a fool's errand, even when attempted via deliberative democracy.[92] Rather, the goal is to decide on *actions* that most people can live with, at least in the short term or if balanced by certain concessions. The goal of getting people with vastly different expertise, understandings of nature, and tolerance for risk together and having them talk with one another is a better understanding of the grounds for disagreement. The point is to discern possibilities for leveraging dissent into more intelligent policy. Outcomes would be partial victories to as many different groups as possible, converting today's chaotic, often gridlocked, status quo into something more productive. Small, halting steps forward would provide more useful opportunities for collective learning. Recommendations for safeguards might be expanded or reduced, depending on how experiments turn out. Proposed funding priorities could shift in light of research results. In any event, our ability to intelligently cope with the complexities of environmental problems suffers in the absence of this kind of institution.

Pask's thylacine lab remains several steps away from resurrecting the Tasmanian tiger. They've mapped the genomes of the thylacine and its dunnart cousin. But there is a great deal of work left to do to synthesize a living thylacine cell and develop assisted reproductive technologies for marsupials.[93] Whatever role de-extinction plays in the future of biodiversity, the effort provokes us to more seriously consider how we ought to govern radical conservation technologies.

BETWEEN RESTRAINT AND RECKLESSNESS

Like so much start-up technology, resurrected mammoths may end up being more hype than reality. In more measured news features, Church clarifies that they are "de-extincting genes, not species." Colossal's goal is really an elephant–mammoth hybrid, one able to tolerate Siberia's climate and to interbreed with, and thus help preserve, the endangered Asian elephant.[94]

Even if it seems overhyped, de-extinction inspires more people to think about biodiversity conservation. We might draw a parallel to electric cars. Elon Musk's first vehicles with Tesla were luxury, niche products, but they also helped drag electrics into the public imagination. The prospect of resurrecting mammoths is similar. For better or worse, it captivates more people than conserving lesser-known and less-charismatic species.

The radical conservation efforts in this chapter arguably do the most good as conversation starters. They provoke us to reconsider what matters to us about nature. As technological developments enhance our ability to intervene in ecosystems and to safeguard species' future, we're faced with difficult questions. Is it really the "naturalness" of coral or the nativity of species that matters to us? Might we better steward the environment by sometimes jettisoning these ideas to reengineer ecosystems for resilience? At the same time, can bioengineers be trusted?

My argument has been that we can't really answer these questions without experience. Our own lives offer helpful analogies. For instance, all the premarriage counseling in the world doesn't totally prepare young lovers for future arguments over how to load the dishwasher or the far more serious disagreements that emerge when raising children or when experiencing financial hardships or serious illness. Likewise, we can't know if radical biodiversity conservation is worthwhile without getting into the thick of it. We may find that engineered coral impart as much awe and support just as many fish as "natural" ones. We might learn to love at least some "out-of-place" species.

Biodiversity conservation, however, is traditionally biased toward precaution. An article in *Conservation Science and Practice* refers to this as the discipline's "ethos of restraint." In light of the large gaps in knowledge, as well as the uncertainties regarding how to contain radical conservation experiments, many scientists advocate waiting. We are to refrain from acting until scientific understanding can eventually catch up and prove these technologies' safety.[95] Political scientist Aaron Wildavsky described this ultra-precautious mindset as calling for "trials without errors," which essentially demands that we do not have trials at all.[96]

Precaution comes with its own costs. As coral geneticist Line Bay put it, "The worst thing that we could do is ignore the genetic engineering because it's frightening for some people, and then get 10 or 15 years down the road and realize it's the only option."[97] Inherited ideas about ecological risk can threaten species' survival. Some scientists argue for preventing the hybridization of barred tiger salamanders with an endangered version endemic to California. Although the offspring are usually more resilient, they seem to have slightly different effects on the ecosystem. Prioritizing the "genetic integrity" of California salamanders, or the previous ecological baseline, risks leaving the local landscape with no salamanders at all.[98]

My point, of course, isn't to call for blindly plunging forward. Intelligent trials of de-extinction, assisted migration, assisted evolution, and gene drives would keep the risks to a minimum while allowing us the ability to learn more about their practicality and consequences. Start small. Build up slowly. Monitor carefully. Transform criticism into reasonable precautions. Protect against conflicts of interests. Take steps to lessen the potential impact on the victims of error—human and nonhuman. Such commonsensical strategies can turn a proposal that initially seems like opening Pandora's box into a mechanism for improving the safety of radical conservation interventions.

By doing so, my hope is that conservation can eventually become more democratic. Biodiversity impacts nearly all of us. Most people have a stake in the question concerning how humanity can best dwell alongside other species. Conservation decisions shouldn't only be made by those who claim to speak for "the best available science," who can most effectively leverage the Endangered Species Act to their side's advantage, or who just so happen to sit on millions of dollars in venture capital funding. Biodiversity democracy would give almost everyone at least some input into how we try to achieve a wilder and more vibrant Anthropocene.

8 THE FUTURE OF BIODIVERSITY

Hot, dry, and sparsely populated, the Little Sandy Desert is a lot like the rest of the Australian outback. Average yearly rainfall is a mere fourteen inches, roughly on par with the aridity of Morocco or Afghanistan. Scraggly acacia trees and spinifex (a spiny, hard grass that cattle usually avoid eating) dominate the landscape. It is here that the Martu people made their home for millennia, until the area was turned into a testing range for intercontinental ballistic missiles.[1]

The Martu resettled the Little Sandy Desert decades later. They returned to a very different landscape. No longer subjected to periodic prescribed burns, the area suffered larger and more intense fires. Native wild species populations had shrunk, despite the ending of hunting. The Martu had changed too. They returned a more sedentary, less nomadic people. They relied far more on imported foodstuffs and ate far fewer of the local species, especially the native grasses and seeds that were labor-intensive to forage. But what intrigued scientists was that a much more domesticated human population seemed to produce an environment that was far less wild.[2]

The Martu's dietary changes don't completely explain the new Little Sandy Desert. Nevertheless, anthropologists argue that the shift to a more specialized diet, largely relying on imported foodstuffs, deprived the landscape of a key ecosystem function. The Martu people's traditional hunting and foraging activities involved a wide variety of species. They ate predators, which reduced pressure on traditional prey populations. They burned the landscape to facilitate hunting. Although environmentally disruptive, this practice fostered a mosaic landscape favorable to high levels of biodiversity. The Martu's diverse culinary tastes actually helped complexify the local ecology. The researchers thus described them as "knitters" of the ecosystem.[3]

My point with this example is not that Indigenous Australians are intrinsically more environmentally ethical than the rest of us. Nor am I arguing that we ought to return to foraging. Rather, the example provides an illustration of another way to think about people's place within nature, one that is neither dismal nor pines for harmony. Through experiments in living within a harsh landscape, the Martu learned to add to the vibrancy of their local landscape while simultaneously taking from it. They constructed a novel ecosystem, an artificial nature that better suited their needs. Their presence made a fragile desert wilder than it would have been without them. We too could better "knit" together local ecosystems, albeit in the context of current technological societies.[4]

Many of the biodiversity reconciliation efforts that I have described are not totally different from the Martu's knitting of nature. Agroecology ties a variety of farm species together while still producing foodstuffs for people. Green roofs shelter humans at the same time they knit together pollinating insects with birds and flowers. And the effort to reintroduce Siberian megafauna could weave together a livelier artic. The difference between contemporary hopes to eke out a sustainable, but modern, lifestyle and the Martu's efforts to survive the outback is really only the scale, stakes, and complexity of the challenge.

I began writing this book after the United Nation's announcement of the Kunming-Montreal Global Biodiversity Framework. The agreement targets massive increases in protected areas and land restoration, alongside reductions in non-native species introductions. It calls for us to live "in balance and in harmony with Mother Earth."[5] What I have tried to show in this book is that hope for achieving something like that will require leaving behind the very political thinking that led to the Kunming-Montreal agreement. Despite allusions to "different value systems," the document really only speaks to a narrow segment of the world's population. It focuses on top-down setting of quantitative targets, it prioritizes protected conservation areas, and it insists on pushing through "transformative changes." The agreement is steeped in twentieth-century environmental technocracy and its associated pathologies.

The foundations for the environmental narratives of the past have crumbled. The start of the Anthropocene arguably predates the Industrial Revolution by millennia, perhaps commencing when early humans started making fire or became too good at hunting megafauna. National parks and other protected areas are artificial creations, made to realize romantic ideals of nature. At the same time, novel ecosystems can be more vibrant than historical ones. Genetic engineering may be the only realistic lifeline for many highly valued species. When modernity no longer serves as a foil for ecological harmony, we are left adrift in the search for seemingly authoritative grounds for defining humanity's rightful place in the environment.

Science journalist Elizabeth Kolbert wrestles with this quandary in *Under a White Sky*. Her book's title refers to a world where the horizon is colored by particles released to control the climate, an allusion to humanity's seemingly insatiable tendency toward environmental tinkering.

She begins with the invasion threat that Asian carp pose to Lake Michigan, a situation made possible by the Chicago Sanitary and Ship Canal. The waterway was originally built to direct Chicagoans' sewer waste away from the lake and into the Illinois River. Now, it risks facilitating ecological contamination. The effort to keep the carp out has culminated in lining a section of the canal with electrical wires, whose shocks should stop the fish from straying. The case illustrates, as much in Chicago as elsewhere, that going back to a pristine nature is no longer an option. Our choice regarding the environment "is not between what was and what is, but between what is and what will be."[6]

Kolbert's book reads as a lament. She presents humanity's ongoing environmental tinkering as ultimately resulting in a diminished version of nature. The scientists working to save the environment with radical technologies act more in the spirit of techno-fatalism than techno-optimism. The outcomes are consolations because restoration is now impossible.[7] Similar sentiments are expressed by the narrator of Kurt Vonnegut's *Galápagos*: "People wouldn't have had to behave so resourcefully, wouldn't have been in such complicated difficulties, if the planet hadn't been made virtually uninhabitable by the creations and activities of other people's great big brains."[8]

Breakthrough Institute founder Ted Nordhaus chides Kolbert for focusing "almost exclusively on all that goes wrong when we attempt to control nature, even as [she] acknowledges that we often have no choice." In Nordhaus's view, Kolbert remains trapped within the eco-apocalyptic narrative, which sees humans as having an invariably pernicious influence on nature. Nordhaus redirects our attention toward hard-fought environmental success stories. We now use less land to grow our food, which allows for the return of forests. Generalist species, such as deer and mountain lions, are thriving.[9]

Yet, Nordhaus seems to fall prey to the fallacy of the excluded middle. There's no doubt that an Asian carp infestation of Lake Michigan would be lamentable, he reasons, but it is preferable to "open sewers running through [the] streets."[10] Although Kolbert arguably portrays human tinkering as too often leading us on the path to ruin, Nordhaus too readily writes off past mistakes as the price of progress. But relentless optimism can lead us astray as much as dismal pessimism can. The ability for technological advancements to assure a stable ecological base for human flourishing isn't guaranteed. Rejecting *Homo extinctor* shouldn't mean uncritically embracing *Homo faber*.

Cases such as the Martu alert us to other possibilities for imagining people's place within nature, so long as we don't romanticize them. The Martu people neither preserved nature nor absentmindedly tinkered with it. Their relationship with the outback appears to have been the result of millennia of searching for win-win relationships. But I wouldn't exactly call it harmonious, for they worked to craft an ecosystem that suited their needs. The Martu encouraged the surrounding wilderness to generate ever-larger surpluses for themselves and the species they depended upon. It is a strategy we could more often mimic.

Yesterday's Chicagoans weren't really faced with a Hobson's choice: build a canal or wallow in shit. If decision-makers at the time had the imagination to compost or bio-digest their waste rather than just flush it down sewer pipes, they could have saved future citizens from a potential biodiversity problem—or perhaps, just as accurately, a sport fishing threat. Illinois conservationists go to war against Asian carp today because yesterday's city managers failed to commune with bacteria. The majority of America's municipal waste solids still gets incinerated or landfilled rather than turned into fertilizer, which illustrates how neglected opportunities for knitting nature often are.[11]

Perhaps that asks too much from nineteenth-century people. Chemical waste treatment was in its infancy. City managers' reference point for recycling sewage would have been its unsanitary transport by horse-drawn carts, which was often applied to fields in ignorance of the bacterial and viral risks.[12] However, the era's barriers in knowledge, technology, and logistics were not insurmountable. It was imagination that was lacking.

Many contemporary problems are solved with a similar degree of ecological shortsightedness. We too often fail to realize potential win-wins that would seem obvious in hindsight. It is a lack of productive disagreement and democratic decision-making that holds us back. Despite the urgency of global environmental challenges, we take too little advantage of opportunities to better knit together more vibrant natural environments.

THE PROPER PLACE OF SCIENCE

For all its splendid discoveries, modern science is the product of a gross philosophical blunder. So argues philosopher of science Nicholas Maxwell. In his 1984 book, *From Knowledge to Wisdom,* Maxwell justifies his dismal assessment by observing how the world's problems only seem to have grown in stride with our scientific capacities. Although hunger is on the decline and new lifesaving medicines abound, climate change and environmental harms present new challenges. Science and technology have given us safer automobiles and airplanes but contribute equally well to the horrors of modern warfare and terrorism.[13]

To Maxwell, this inconsistent relationship between science and human progress can be traced back to the mistaken belief that the fundamental purpose of science is to pursue knowledge rather than wisdom. We have tended to believe that if we discover scientific truths, then the resolution of humanity's problems will no doubt follow. When it comes to biodiversity, surely science can tell us, then, whether we should degrow our economies and return to quasi-pastoral village life or, alternatively, churn out nuclear power plants and admire a pristine but largely off-limits wilderness from our high-rise apartment's balcony.

In Maxwell's view, this mindset is irrational, even neurotic. Instead, scientists ought to have been helping us answer questions more directly tied to human well-being. What do we value in life? What problems do face in trying to realize those values? And what are solutions to those problems? Although knowledge is doubtless an essential component of doing so, the more fundamental challenge is one of *action*. Unraveling humanity's most pressing problems is not a mere matter of knowing more but rather of better learning how to cooperatively solve them.[14]

Some readers might object, arguing that science is perfectly fine. Humanity's problems only seemed to have grown alongside our knowledge about reality because politicians have failed to heed the voice of experts.

Maxwell disagrees. He argues that it is not the job of scientists to authoritatively tell us what should matter to us, what we should consider our problems to be, and how we ought to solve them. Their job is "to help the rest of us reach our own decisions."[15] I hope that if have made one thing clear in this book, it is that Maxwell is right. The biases of the Living Planet Index show that we should be wary of outsourcing our level of panic to seemingly objective models. Likewise, whether the Colombia hippo is invasive and whether we ought to cull the population as a result are not really questions that can be resolved scientifically. It depends on how we think about nature, with which species we prefer to ally ourselves, and how hard we're willing to try to nonviolently reconcile interspecies conflicts.

In short, science is woefully incapable of fulfilling the political role being asked of it. Even worse, demanding that science arbitrate environmental disputes often backfires. Recall the Netherland's state nitrogen model that has become excessively authoritative in Dutch political decisions. The electoral successes of the populist farmers' party and far-right politician Geert Wilders are partly a result of their promises to free citizens from technocratic environmental administration run amok.

I was reminded of Maxwell's arguments while attending the plenary meeting of IBPES. The meeting itself was nothing to get excited about. I looked on as national representatives endlessly debated how they should word a statement about invasive alien species. It felt like a weeklong faculty senate meeting, which is to say that the people who voluntarily subject themselves to the process are saints. God bless them.

It wasn't just the drudgery of the process, however, that stood out to me. As someone who researches politics, I was surprised by how limited the disagreements were. Disputes, of course, did happen. European representatives insisted on emphasizing "precaution" wherever possible. The Bolivians pushed for language that endorsed Indigenous knowledge, whereas African participants strove to refocus attention on the impact of ecological pests on human well-being. I have no doubts that the resulting consensus document helps national politicians and bureaucrats advocate for local conservation policies. But I was disappointed that important definitions, and the tone and thrust of the document, were already settled. Most of the problems with the concept of alien invasive species, which I surveyed in chapter 3, weren't up for discussion. What would more heterodox biologists, such as Chris Thomas, think about IPBES's scientific summaries? Would compassionate conservationists, who oppose the mass killing of animals in the name of biodiversity, consider the report to be so authoritative? Where was the influence of Indigenous groups who see invading species as best controlled by being used rather than eliminated?

At the IPBES meeting, I saw that the linear model of science policy thinking is alive and well. The very act of creating grand scientific assessment documents is guided by the idea that amassing research data will invariably spur political action. It is hard to shrug off the dreams of conservation technocracy. Many of us desperately want to believe that collecting measures, models, and maps of the extinction threat, the cost of biological invasion, and the benefits of ecosystem services will provide an unbiased and stable base for policymaking.[16]

This book has been about the challenge not only of imagining a positive place for humans within the environment but also of discerning the proper role of science within democratic societies. And to be clear, my contention isn't with scientists so much as with the unfair burden that has been placed upon their shoulders. As Daniel Sarewitz argues, we often ask science "to do the impossible," to deliver "scientifically coherent and politically unifying understandings of problems that are inherently open, indeterminate and contested."[17]

When science gets overextended into democracy's turf, it is usually not scientists' fault. Citizens, journalists, and powerful politicians simultaneously demand too much and not the right things from science. Environmentalists try to make biodiversity into a moral lodestar. Competing partisans try to leverage competing facts into political advantage. Governments point to experts in order to abdicate responsibility, to act as if their hands were tied when implementing contentious policies.

To be fair to the good people involved with IPBES, they are currently doing a far better job than the IPCC has, at least in terms of recognizing the value ladenness of expert guidance. The organization's science advice will eventually incorporate values through a framework they call "Nature Futures." This framework acknowledges that people cherish the environment for a variety of reasons. Preservationists value wild nature for its own sake, referred to as "nature for nature." The fact that people rely on the environment for food, fiber, and other ecosystem services is labeled "nature for society." And the "nature as culture" pillar reminds us that many people's traditions and faiths are bound up with the environment.[18] Nature Futures is a step forward in being more honest regarding how different values lead to competing visions of a good Anthropocene.

But something feels wrong about converting political differences into inputs for scientific modeling. Although the Nature Futures framework is more inclusive than traditional conservationism, it still feels like an effort to rationalize politics. One senses the hope that the messy unpredictability of biodiversity conflicts can be quieted so that they can be managed through more enlightened bureaucratic administration. Start with several cups of nature for society. Mix in a few quarts of nature for nature, and sprinkle generously with nature as culture. Armed with fine-grain environmental valuation surveys, could we arrive at less contentious and conflictual environmental policies?

I am wary the very idea of having scientists try to imagine our futures for us. I don't want IPBES to provide us with models for living in harmony with nature, even if they are built upon detailed public surveys. I question the hope that synthesizing thousands of pages of scientific research can reveal unambiguous pathways toward unifying and transformative socioenvironmental changes. In my view, experts are better used as bridge builders than as prophets.

This book has surveyed much of what divides citizens when it comes to biodiversity. The Old West versus the New West. Rural peoples versus fortress conservationists. Rewilders versus invasion biologists. Doomsayers versus skeptics. Corporate-friendly new conservationists versus old-guard preservationists. I'm not sure these disparate groups really need their own environmental values fed back to them in the form of ecologically harmonious scenarios, as if the right vision of the future will finally deliver the consensus that has eluded environmental politics for more than a century. Instead, we need experts to help us better navigate the turbulent waves of conflict, to find hidden currents that bring us through swirling eddies of contention. We need them to help us knit together never-ending disagreements into tentative collective decisions.

Step 1 would be to demand a tighter connection between scientific research and practical problem-solving. Redirect a fraction of the awesome brain power now being used to ever more minutely catalogue hazards toward discovering solutions. A researcher working in archaeological food web modeling, the method used to study the Martu, once came to give a talk at my university. They presented evidence of food webs becoming simplified whenever humans have settled an area. When I asked what lessons their research could offer us in partly reversing these changes, they looked as if they had never seriously considered the question. That is a problem.

Step 2: Describe scientific research in terms of the values and interests they serve (or challenge). Studies that turned preliminary evidence of the Colombian hippos' potential environmental impacts into a seemingly self-evident call for their removal were obviously preservationist in orientation. That such research is value laden isn't a problem per se. All research is unavoidably biased, rooted in subjective judgment calls that scientists must make in deciding what to study, how to study it, and how to interpret the results. However, *not acknowledging* that bias, that one's work is guided by a particular story about nature, is a problem.

Finally, step 3, scientists (and funding agencies) could more often ask, "What research would better help solve the biodiversity problems that people care about and in ways that help them reconsider and, hopefully, partly reconcile their diverse values?" There are many answers to that question. It might mean doing more studies on making meat eating more compatible with biodiverse grasslands, more experiments on reintroducing previously extinct animals, or finding better nonviolent means to control non-native species. Who knows? The point is that we demand that research tries not just to describe the world but also to help us collectively arrive at better decisions, especially by discerning *actions* or *solutions* that could make political compromises easier to achieve.

A greater portion of citizens ought to have at least some influence on major conservation decisions. A sensible democratic society would create venues where concerned citizens could more reliably signal the biodiversity issues that worry them as well as the types of research and solutions that they believe deserve more attention. It would support a class of publicly engaged scholars and journalists. Their job would be to put scientific findings into the context of long-lasting disagreements, highlight and advocate for avenues of research that could partly resolve environmental disputes, uncover barriers to action, and push people to rethink simplistic environmental narratives.[19] The goal would be to foster more productive public discord. Where "follow the science" has failed, biodiversity democracy can succeed.

ENVIRONMENTALISM FOR EVERYONE

> [There is an] absolute necessity of waging all-out war against the debauching of the environment . . . The bulldozer mentality of the past is a luxury we can no longer afford. Our roads and other public projects must be planned to prevent the destruction of scenic resources and to avoid needlessly upsetting the ecological balance.[20]

Who said this? Ralph Nader? Al Gore? Bill McKibben? Actually, it was Ronald Reagan, in 1970, when he was still governor of California.

President Reagan's relationship with the environment is not my focus here. Let's just say it was complicated. He oversaw an effort to decimate federal environmental agencies. Yet, he also signed the Montreal Protocol, which phased out ozone-depleting chemicals.[21]

My point is to return to the question of why environmentalism, especially in the United States, has become so polarized—and how it could be otherwise. I've already mentioned how rural livelihoods were increasingly put in conflict with protecting nature over the-twentieth century. This was a by-product of the design and implementation of environmental law, which threatened to exacerbate the already-declining economic prospects for Old West livelihoods. Perhaps most damaging is the idea that there's only one way to be an environmentalist.

In his book, *Getting to Green*, Frederic Rich charts how protecting the environment became central to the divide between Democrats and Republicans.[22] The American right leveraged the growing conflict between environmental regulations and economic livelihoods for electoral gains. Leftist environmentalists increasingly embraced eco-centric philosophies, ideologies that often portray saving trees and protecting rare species as more important than human beings. The result is that environmentalism ceased to be a bipartisan issue. Instead, it feeds a never-ending culture war.

The role of the environment in partisan signaling, however, is only part of the picture. Protecting nature divides us not just because of the aloof anti-humanism of some environmentalists and the political opportunism of right-wing populists. People are also motivated by differing cultural theories. Recall the four categories that I describe in chapter 3: egalitarian, hierarchist, individualist, and fatalist. Egalitarians worry primarily about inequality. They also think that nature is fragile and wish for everyone to collectively embrace less material-intensive lifestyles. Eat vegan. Tread lightly. Live simply. On the other hand, nature to a hierarchist is something to be managed in order to keep it within acceptable limits. They trust experts to administer the problem, to know how everything ought to fit together in order to ensure a healthy environment. Individualists tend to see nature as inherently benevolent, or at least that human ingenuity and market competition can avert catastrophe. There may not be much point talking about fatalists, since they see nature as capricious and human influence as trivial.

These four cultural theories map more or less onto different biodiversity debates. For an issue such as alien species, worrying about disrupting nature's balance puts egalitarians in coalition with hierarchists. The latter view newly arriving plants, animals, insects, and other living beings as out of place, as ecological rule breakers. Individualism, in contrast, is visible in Chris Thomas's recasting of species as either successful or not rather than as invasive alien or embattled native. Individualism hybridizes with egalitarianism in compassionate conservationism, which views even invasive alien species as having some right to make a home in the landscape.

Cultural theory helps us see that it rarely makes sense to divide the world into pro- and anti-environmentalists. Rather, there are competing and sometimes disparate *environmentalisms*. The denial of this diversity drives polarization, converting normal political conflict into outright nastiness.

The trouble is that we struggle to think outside of our own preferred cultural theory for an issue. Its assumed scientific or moral correctness appears indisputable to us, which makes anyone who disagrees into a science denialist or even worse. Writing in 1992, entomologist Ralph Wills Flowers portrayed opponents of the Endangered Species Act as promoters of a "new final solutionism," effectively the moral equivalent of Nazis.[23] A similar inability to see disagreement as legitimate infects the debate between advocates of degrowth, a plan to purposefully reduce economic activity to keep humanity within planetary limits, and self-described eco-modernists, who believe that technological progress can provide a path to a good Anthropocene.[24] That the two sides talk past each other is unsurprising because each is based in a competing cultural theory.

The resulting deficit in political curiosity and imagination leaves us locked into polarized environmental politics. Consider again the Kunming-Montreal agreement. Mirroring earlier climate protocols, fulfilling international biodiversity agreements will demand cumbersome monitoring systems to audit protected areas and restored lands. Nations are essentially asked to centrally plan their landscapes for increased biodiversity. The agreement recommends an unavoidably antagonistic approach to tracking and evaluating companies' biodiversity impacts.

In short, the agreement is a hierarchist's dream. The fuzziness of biodiversity, however, will make it hard to transform the idea into something to bureaucratically administrate. How "pristine" must a protected area be? What exactly does "restored" mean? Compared to what? When exactly is a species non-native or invasive? What is "sustainable" agriculture?

In any case, should the Kunming-Montreal agreement actually spur national action, it will be divisive. Individualists will reject the impetus toward new restrictions and regulations. Egalitarians will worry about conservation evictions, despite language about working to earn local communities' "informed consent."

We've been here already. Writing in 2006, political scientist Marco Verweij wondered aloud whether the Kyoto Climate Protocol was "merely irrelevant" or "positively harmful" when it came to combating climate change. He argued against convoluted hierarchist schemes such as carbon credits or emissions trading. Instead, Verweij advocated directing R&D to discover alternatives to fossil-fuel technologies in energy, transport, and elsewhere and to make them as cheap as possible. We would then assist developing countries to create their own R&D centers and build up renewable infrastructures.[25]

Verweij's proposal is, at least to the first approximation, what has come to pass. Spurred on by government support and technological innovation, renewables are cheaper than coal. And electric vehicles are poised to eventually overtake the internal combustion engine. There is some, although not overwhelming, evidence that agreements such as the Kyoto Protocol haven't helped. The rate of decarbonization has recently slowed in signatory nations, perhaps due to policy debates getting redirected toward targets, credits, and emission restrictions and away from directly supporting innovation and infrastructural changes.[26]

Addressing climate change via larger-scale technological changes is an example of what policy scholars call a "clumsy solution."[27] These are policies that explicitly try to incorporate elements of individualism, egalitarianism, and hierarchism, even if the result seems messy, insufficient, inelegant, or even contradictory. By focusing on making fossil-fuel alternatives cheaper, individualists can extol the unleashing of entrepreneurial initiative.

Egalitarians would see developing nations benefiting from technology transfer agreements. Adopting stricter environmental standards for industry, electrical systems, and consumer products after corresponding technological advancements can help hierarchists trust that the transition is being appropriately managed by experts. Clumsiness is a strategy for helping nearly everyone claim a (partial) moral victory from a decision, even if they aren't entirely happy about it.

Some of the win-wins that we have looked at so far have been clumsier biodiversity solutions. To be fair, the Kunming-Montreal agreement mentions a few of them. Biodiversity offsets combine the desire of egalitarians to protect fragile nature via protected areas with the individualists' interest in continuing to permit economic activity. But as we saw for Guinea, they require hierarchical monitoring, rules, and sanctions for violations to function well. Indeed, many of the reconciliation efforts from that chapter could be criticized for still relying too heavily on the existence of powerful environmental bureaucracies. What would biodiversity solutions that prioritized the cultural theories of individualists and egalitarians look like?

An overlooked clumsy conservation solution is simply focusing more R&D on developing biodiversity-friendlier technologies. For instance, rather than implement difficult-to-govern biodiversity offsets for mines, might we try harder to make mining less disruptive (e.g., less noisy, less polluting) to surrounding wildlife or improve our ability to recycle valuable metals? Research dollars could be earmarked for civil engineering faculty to more often collaborate with ecologists. How might we better foster biodiversity in non-pristine places? Could we more cheaply build wildlife road crossings or better shelter wildlife within cities while simultaneously reducing the likelihood of human–animal conflicts? Decreasing the cost of and improving standards for bird-safe glass could prevent hundreds of millions of birds from fatally colliding with windows. Freshwater biodiversity might bounce back more quickly by dedicating more research into less environmentally impactful hydroelectric dams rather than just calling for their demolition.

Likewise, we could better balance agricultural needs with biodiversity. For instance, changing when rice farmers flood their fields can provide temporary wetland habitat for migratory birds. According to one University of Missouri ecologist, committing a mere 10 percent of farmland to strips of wild prairie could reduce fertilizer runoff and soil erosion by more than 90 percent and bring big bumps in insect and bird biodiversity.[28] More research might uncover still more possibilities for making agroecological ideas more productive, less labor-intensive, and cheaper to implement.

The opportunities for reconciliation are endless. Small design changes can make solar farms compatible with grazing and the planting of shade-loving crops. Some researchers have found that the land beneath panels can be used to create communities of lichen, bacteria, and mosses that are used for desert land restoration.[29] Conservationists have developed "beaver dam analogues" that can either support reintroductions or spur wetlands to develop where the animal can't yet return. And still others have come up with "beaver flow devices," which reduce the likelihood of flooding farmers' fields by giving excess water an escape route.[30]

The above technologies and still others could be developed in ways that make them attractive to the developing nations that host most of the world's biodiversity hot spots. However, compared to calls for more protected areas, these approaches suffer from having too few passionate advocates.

Some conservationists, however, already recognize the need for biodiversity to be clumsier. Activists in Scotland, for instance, increasingly eschew the traditional storyline of rewilding. No longer do they emphasize predator introductions and landscapes without people. Instead, they insist upon the compatibility of rural people and rewilding as well as the potential for transforming rural economies. And they put less stress on the need for controversial species reintroductions, at least for now.[31]

The fact that rewilding advocates in Scotland have realized the importance of political clumsiness gives me hope that the same can happen elsewhere. It will take pushing back against a long history of purism in environmental politics, as well as against the broader pattern of political polarization in the United States and elsewhere.[32] Conservation need not be at the center of future culture wars. But that demands a more patient and humble kind of environmental politics, one that speaks to environmentalists of all stripes.

LETTING GO OF ENVIRONMENTAL RATIONALISM

I grew up in a working-class suburban family, which is to say that my exposure to nature was limited to what bugs and animals could be found at recess and in the backyard. I was first exposed to a whole new world when, as a young boy, my grandparents took me to Rocky Mountain National Park. Spending extended periods of time outside of a city, however, only came when I moved to New Mexico. I hiked the local peaks, went mountain biking, and backpacked in the Gila Wilderness. I read Edward Abbey and Daniel Quinn, largely embracing the preservationist-oriented environmental philosophy that is typical of people who go on to enroll in PhD programs. I began thinking more seriously about biodiversity after reading George Monbiot's book, *Feral.* At the time, I was enthralled with the idea of rewilding. I would daydream about rhinos and camels roaming the New Mexico scrublands, a landscape that now seemed to fall short of its ecological potential.

Although my love of nature hasn't changed, my view of the politics of biodiversity conservation certainly has. I am embarrassed to admit that it took writing a book for me to realize how dogmatic my thinking was, to see how I uncritically embraced a kind of environmental chauvinism. I had little respect for people who made their living from the land, who produced the very goods that I used to live a comfortable life as a middle-class academic. I bought into the idea that every blade of grass eaten by a dairy cow or square foot of earth settled by people was a sin against nature. I thought the case for an impending ecological crisis was airtight, that the necessary changes only lagged due to a failure in political will, a stubborn resistance to doing what was obviously right.

That is not to say that I think everything is okay with the world's ecosystems. Species are going extinct, to uncertain effect. Many fisheries are in decline. Coral reefs are disrupted by high temperatures and ocean acidity. Climate change will challenge us in innumerable ways. Yet, I now also recognize that many things are going well. Tigers are thriving in India. Populations of bison, beaver, bear, wolf, and countless other species have dramatically rebounded since 1960. Even though deforestation in Brazil and elsewhere remains concerning, tree coverage has expanded across North America, Europe, and Asia. Humpback whale populations are nearly back to where they were before the species was nearly hunted out of existence.[33] Success stories get overlooked because of the widespread bias toward environmental pessimism. In any case, taking a more uncertain outlook, rather than a catastrophic one, has opened my eyes to a wealth of opportunities that lie before us. I can no longer make sense of the dismal perspective that views large-scale human civilization as inevitably driving toward to an ecological apocalypse.

Don't get me wrong, there's a chance we might be doomed. But I have become convinced that alarmism isn't helpful for dealing with humanity's global challenges, even if we could succeed in getting nearly everyone sufficiently frightened. There have been periods of time defined by environmental consensus, when widespread agreement regarding ecological decline enabled seemingly unstoppable political progress. The 1960s and 1970s in the United States was such a time. However, *too much agreement* in that era set us up for the dysfunctional political conflicts that followed. A lack of dissent allowed mid-twentieth-century politicians to act thoughtlessly. It let them pass environmental laws that were too big, that were too radical, inadvertently creating the next generation's conservation enemies. As understandable as it is to pine for the golden age of American environmental legislation, policy revolutions typically set the stage for future backlashes.

I am not a conservative myself, but I still find philosopher Michael Oakeshott's words useful for thinking about the limits of twentieth-century environmentalism. I am reminded of his essay, "On Being Conservative," whenever I hear calls for humanity to live in harmony with nature or demands for transformative change. Oakeshott warned about political rationalists, those who yearned to establish order and coherence to the chaos and untidiness of the human world.[34] Rationalistic environmentalism pines for a planet where conflict no longer exists, where the relationships not only between humans and nature but also between people are harmonious. It is a world in near complete consensus regarding the merits of ecological living. Experts provide comprehensively correct answers regarding how to dwell in nature, and governments need only be willing to authoritatively impose those designs.

Indeed, some environmental scientists openly yearn for global institutions with the authority to force nations "to ensure that . . . planetary boundaries are respected." CBD head Kierán Suckling hopes for a future with a Trump-like, albeit preservation-minded, president. He envisions someone willing to use executive decrees, guided by "the best available science," to aggressively conserve nature.[35] As Oakeshott wrote, political rationalists act as if they believe that "to govern is to turn a private dream into a public and compulsory manner of living."[36]

Scientific expertise, however, is frustratingly inadequate when trying to build comprehensive designs of harmonious futures, ecological or otherwise. The uncertainties and complexities loom too large. Biased assumptions, subjective values, and simplifying stories invariably plug up the gaping holes in our collective knowledge. This is why I end this book with a call for us to let go, at least a little, of the rationalistic dreams of twentieth-century environmentalism. We need to focus less, not more, on trying to make Aldo Leopold's "new kind of people" into a reality.

Today's environmentalists err when they present popular disagreement or skepticism about their preferred policies as the product of ignorance or misinformation. It is a mistake to long for, as Canadian writer Graham Saul does, an environmental movement that can "convey truths that cannot be reasonably denied."[37] We need disagreement. Popular dissent provides vital political feedback. It tells us that proposed changes are missing something or risk unintended consequences. We learn that the harmonious dream behind environmental policies speaks to too few citizens' values. Or it may be that the people offering up that dream aren't seen as persuasive or as trustworthy, regardless of whether it involves limits to growth or ecological modernization. A more democratic path toward a wilder planet will require embracing discordance and disagreement as features, not flaws, in the political process.

TOWARD BIODIVERSITY DEMOCRACY

After writing this book, natural landscapes and the world of politics no longer seem so different to me. Ecologists also once believed in a kind of rationalism—namely, that undisturbed nature would find a stable balance.

They thought, for instance, that forests follow a pattern of "succession" to reach a "climax state," what scientists of the era must have seen as a rational image of environmental perfection. However, as Daniel Botkin describes, a new image of nature has emerged: "Change now appears to be intrinsic and natural at many scales of time and space in the biosphere . . . and in at least some cases these changes are necessary for the persistence of life, because life is adapted to them and depends on them."[38] No longer is it expected that fenced-off forests would attain a perfect environmental balance and remain in that state forever. Rather, ecosystems are constantly adapting and evolving in response to a never-ending stream of disturbances. Trees fall, burn, or die out to disease or drought. Forests become grassland and sometimes return. Species populations boom and go bust. Pioneer species invade disturbed areas only to be put back in their place at a later time. Nature is defined, as Botkin puts it, by "discordant harmonies."

There is an emergent intelligence at work in these dynamic discordances. A grassland prairie, for instance, is filled with species that both compete and cooperate in innumerable unconscious ways. When bison roll on the ground to ward off bugs, they create bare patches. These spots become microhabitats for arthropods, a food source for insect-eating birds while plant life tries to reestablish itself.[39] What would seem to be the small-scale butchery of grasses ends up setting the stage for a biological form of creative destruction.

Countless species interact in ecosystems, but rarely is any one of them so powerful as to be able to unilaterally impose their own preferred order. Species mutually adjust to each other's behavior over time, which generally leaves the environment more vibrant and livelier as a whole.

This kind of collaborative dynamism is also found in the social world. We avoid running into one another in a busy intersection without a central authority directing traffic. Pedestrians pick up on subtle cues, making mutual adjustments to their trajectories. An awesomely complicated array of relationships assures that I can buy an afternoon coffee at a café, without there needing to be anyone behind the scenes to turn the economic dials. Cultural exchange, moreover, drives endless linguistic, culinary, and intellectual differentiation and hybridization. Yet, we often struggle to transform the immense intelligence lying dormant within complex human societies into better political decision-making.

If humans do achieve harmony with nature or among each other, it too will be a discordant one. It will be conflictual, defined by dynamic disruptions and ongoing change. Competing ideas about what nature needs, regarding what makes for a vibrant Anthropocene and how to realize it, will not disappear. The task is to better channel them toward solving the problem of how to live together despite these differences.

Rewilding the Anthropocene will be a process of *socio-ecological succession*. The focus of conservation shifts away from pristine parks and toward promoting a planet-wide patchwork of diverse experiments in ecological living. The goal is to help foster mutual readjustments between humans and nonhumans, to adapt nature and ourselves so as to better cope with each other's disruptions. The task for environmentalists is to unceasingly advocate that this mosaic be ever more full of life, whether the land be a parking lot or a national forest. Even a suburban neighborhood could be a livelier ecosystem. Sure, it is right to grieve for the loss of the forest that preceded it. But I worry about what happens when we become consumed by ecological mourning. That can lead us to lose sight of the ever-present potential to weave together an incrementally wilder ecological tapestry, one that includes suburbia and other places that have come to symbolize everything biodiversity is not. Ecological mourning can leave us trapped in the preservationist's dismal tale of humans as ignorantly and inevitably polluting nature's beauty. Rather than continue to dwell on humanity's past environmental sins, I argue that we ought to more often take pride in ongoing efforts to achieve an ever-wilder planet.

This process entails a more unpredictable and less circumscribed search for conservation solutions. There will be disturbances. Some lands will be converted into farm fields or mining pits. Many species will go extinct. Hopefully, we can learn more quickly how to lessen the ecological blow of these disruptions. Yet, at the same time, grazing lands and human settlements will also contract or even be abandoned. Our environmental values and visions of a good Anthropocene will evolve. We might even come to find out that it is not as hard to reconcile something resembling modern society and nonhuman ecosystems as we first thought.

This process is arguably already underway. Different nations and states have divergent environmental regulations and cultivate different local landscapes. We learn about unintended environmental harms produced by the goods we create, the methods we use to extract useful resources, and the way we dispose of our wastes. People do strive to do better, even if sometimes it isn't until they're provoked into action by the protests of environmental activists. However, too often, one vision of nature dominates the conversation, whether it be the narrative of fragility or of an infinitely resilient planet Earth. Sometimes environmental disturbances get too big before humanity finally makes adjustments, whether because of political gridlock or some other reason. We are still learning how to effectively and democratically negotiate environmental politics, to more reliably and sustainably reconcile competing environmentalisms.

Calls for top-down transformation forget that living in a democratic society binds us all together in a journey of collective discovery. It's like a giant arranged marriage. Even though we didn't choose to be together, we still have to find a way to coexist and thrive. Democratic pathways toward living better within nature can, at most, be *inspired by* computer models, scientific theories, and grand environmental philosophies. Our ecological common interests can only be revealed, through a lens darkly, by tentative political agreements. Those agreements will change over time, but not because the exhortations of scientists or environmentalists finally wake people up to natural or moral truths. Today's skeptics will *consent* to more radically ecological futures only by gaining experience with, and thereby learning to trust, ways of life that are friendlier to nonhuman species.[40] Otherwise, there is little hope of convincing them that the trade-offs or inconveniences are worthwhile.

This approach to conservation entails a lot of compromise. And, fair enough, many conservationists will argue that all they do is make concessions to appease industry, farmers, and consumers. If I have erred in the previous chapters, it is probably in writing too little about the losses suffered by the environmental left or about the politically privileged position of business. I felt it more important to break from that dismal narrative, to find where environmentalism's traditional opponents had reasonable grievances, and to avoid writing yet another book that preaches to the environmentalist choir.

It is important, however, to keep in mind that there are different kinds of political losses. Some result from polarized clashes and are ultimately decided by legislative intrigue, judicial decree, bureaucratic fiat, or the machinations of privileged interest groups. Those are the worst kinds of politics. It leaves losers frustrated and embittered, setting the stage for ENGO lawsuits and armed standoffs in the Nevada desert. Democratic compromises, on the other hand, allow most sides in a dispute to secure at least a partial victory. They offer participants hope that today's losses can be partly reversed in the future. They seek to expand the range of people benefiting from change and to convert some of yesterday's opponents into tomorrow's political allies. As political scientist Charles Lindblom put it, they "create . . . opportunities for collective action that did not exist before."[41]

At first sight, democracy can seem to be irresponsibly slow, especially given the urgency of global environmental challenges. Compromises appear to shrink our horizons, or they kick the can down the road in order to avoid making difficult, but seemingly necessary, sacrifices. However, if we get the negotiations right, then successes can snowball, enabling a greater portion of citizens to believe in the desirability and feasibility of ever more ambitious conservation policies. What purists deride as selling out or "getting in bed with the enemy" can actually be a more effective strategy toward a wilder Anthropocene, especially if environmentalists are relentless in demanding new compromises if and when the old ones have proven tolerable.

There is a bright future for biodiversity, but only if conservation decisions *work* for most everybody. That demands that environmentalists see skeptics' reservations, desires, and criticisms as politically legitimate, even if they personally don't agree with them. A radical break with the conservation status quo might eventually be possible. But we'll get there only by small steps, by working to bring nearly all citizens along for the ride and letting them even have a turn at the rudder. We'll get there by collaborating with corporate industries and agribusiness, even if those are uneasy, morally conflicted collaborations. As philosopher Richard Rorty once wrote, "In democratic countries you get things done by compromising your principles in order to form alliances with groups about whom you have grave doubts."[42]

The path toward a wilder Anthropocene will require the rewilding of environmental politics itself. It means rejecting the seductive allure of idealized designs for socio-ecological harmony or transformative change and instead embracing a more emergent, uncertain, and dynamic process of biodiversity democracy. It involves taking steps to steer away from the problems of the present but without knowing exactly where humanity is headed. The way to a wilder world is not something we can philosophically deduce or mathematically calculate but rather an ever-shifting route that we can only discover together. Given our chronically restless species' history of braving open waters, there may be no other task for which we're better suited.

Acknowledgments

It took a lot of other people's money to write this book. I should begin by thanking the Alfred P. Sloan Foundation for their generous grant. Without them, this book wouldn't exist. The manuscript's foundations, however, were laid by a Fulbright-funded stay at the German Centre for Integrative Biodiversity (iDiv) in 2022. Being a "fly on the wall," as scientists researched and debated conservation, was an invaluable experience. I am also grateful for a fellowship from the Käte Hamburger Centre for Apocalyptic and Post-Apocalyptic Studies in 2024, where I finished the manuscript and thought long and hard about whether anything of democratic value could be wrestled out of environmental apocalypticism (I still think "no").

At iDiv, I was blessed to have been surrounded by many smart, hardworking, and fun people. Henrique Pereira was exceedingly hospitable, even putting up my family when a health emergency caused us to miss our flight home. I am grateful for how he pulled back the curtain on biodiversity for me. I treasured discussing science politics with Volker Hahn, and his willingness to listen to my not-quite-fluent German. Countless others offered insights and good conversation: Corey Callaghan, Miguel Fernández, Néstor Fernández, Marek Giergiczny, Jess Hines, Theresa Jörger-Hickfang, Jesse Junker, HyeJin Kim, Roel van Klink, Stefan Knauß, Annett Masur, María Méndez, Laura Quintero, Luiße Quoß, Sandeep Sharma, Josiane Segar, Jose Valdez, Marten Winter, and Annika Zuleger. Thank you to everyone who helped me during my stay.

I should also thank several dozen New Mexico Tech undergraduate students, whose coursework amounted to a trial run through the arguments and case studies found in this book, for their thoughts and insights. I'm also grateful for the editing skills of Jaime Terrazas Lozano.

My former iDiv colleagues will all find at least something to disagree with in this book. Nonetheless, I appreciate their patient efforts to help me understand what biodiversity's deal was. Equally innocent of any endorsement of the arguments made in the few hundred pages presented here is the Alfred P. Sloan Foundation. They only believed that my perspective was worth hearing. The Käte Hamburger Centre, it should be noted, invited someone like me for a research stay, despite the fact that my viewpoint on the apocalypse runs (largely) contrary to their institute's premises.

My own analysis owes a great debt to the works of others, people whose writing helped me to sort out my own thoughts on the subject of biodiversity. Especially important have been the books and articles of science journalists Emma Marris, Elizabeth Kolbert, Keith Kloor, Fred Pearce, and countless others, whose writings can be found in *High Country News*, *Smithsonian Magazine*, *National Geographic*, *Undark*, and other high-quality magazines that deserve far more subscription dollars. Thank you for your hard work interviewing interesting people, reporting on important conservation happenings, and asking difficult questions. Otherwise, I'd have had a much more difficult time synthesizing such a broad range of issues. Heterodox thinkers within the environmental sciences, especially Chris Thomas, Peter Kareiva, and Mark Vellend, also inspired me to dig deeper into the divides within biodiversity.

And, of course, my wife and two small boys were a constant source of joy and contentment (and occasional frustration) during the writing process. Thanks for keeping me balanced! I also thank the production team at the MIT Press, including Katie Helke, Justin Kehoe, and many others, whose hard work helped make this book a reality.

Every time I write a book, I wonder if it will be my last. Given the possibility that this one really is it, I want to dedicate it to Ned Woodhouse, Michael Bouchey, and Colin Garvey. We all share not only an intellectual perspective but also, and more importantly, a common commitment to doing social analysis that can matter to ordinary people. I hope I did us proud.

Notes

CHAPTER 1

1. Christopher Ketcham, *This Land: How Cowboys, Capitalism, and Corruption Are Ruining the American West* (New York: Penguin Random House, 2019), 45–48; and Tay Wiles, "How the Feds Helped Make Cliven Bundy a Celebrity," *High Country News*, April 30, 2018, https://www.hcn.org/issues/50.7/sagebrush-rebellion-celebrity-scofflaw.
2. National Parks Gallery, "Desert Tortoise Enjoys a Prickly Pear Cactus Fruit," *PICRYL*, 2010, https://picryl.com/media/desert-tortoise-enjoys-a-prickly-pear-cactus-fruit-89f9c6.
3. Ketcham, *This Land*; and Wiles, "How the Feds Helped Make Cliven Bundy."
4. Ketcham, *This Land*, 48.
5. Ketcham, *This Land*, 48; and Christopher Ketcham, "Cliven Bundy and His Cows Continue to Flout the Law," *New York Times*, September 27, 2023, https://www.nytimes.com/2023/09/27/opinion/bundy-battle-federal-government.html.
6. Greg Eckert and Glenn E. Plumb, "The Language of Biodiversity: A Glossary," *Park Science* 31, no. 1 (2014): 8–10.
7. Mark Magnier, "UN Officials Warn that the World Is in Bad Shape," *South China Morning Post*, May 22, 2021, https://www.scmp.com/news/china/diplomacy/article/3134451/promoting-chinas-biodiversity-conference-un-officials-warn.
8. United Nations Environment Program, "From Agreement to Action: The Road Ahead for the Global Biodiversity Framework," May 20, 2023, https://www.unep.org/news-and-stories/story/agreement-action-road-ahead-global-biodiversity-framework.
9. United Nations, "UN Report: Nature's Dangerous Decline 'Unprecedented'; Species Extinction Rates 'Accelerating,'" *Sustainable Development Goals*, accessed December 13, 2023, https://www.un.org/sustainabledevelopment/blog/2019/05/nature-decline-unprecedented-report/; and Katherine Richardson, Will Steffen, Wolfgang Lucht, Jørgen Bendtsen, Sarah E. Cornell, Jonathan F. Donges et al., "Earth Beyond Six of Nine Planetary Boundaries," *Science Advances* 9, no. 37 (2023): eadh2458.

10. Gloria Dickie, "Explainer: Why Did Past Targets to Protect Nature Fail over the Last Decade?," *Reuters*, December 8, 2022, https://www.reuters.com/business/environment/why-did-past-targets-protect-nature-fail-over-last-decade-2022-12-09/.

11. Convention on Biological Diversity, "Decision Adopted by the Conference of the Parties to the Convention on Biological Diversity," October 15, 2021, https://www.cbd.int/doc/decisions/cop-15/cop-15-dec-01-en.pdf.

12. Laure Fillon, "More Protected Areas Won't Save Biodiversity, Warn Experts," *Phys.org*, January 19, 2022, https://phys.org/news/2022-01-areas-wont-biodiversity-experts.html.

13. Ajit Niranjan, "Food 'Key Driver' for Wildlife Loss," *DW*, March 2, 2021, https://www.dw.com/en/farming-food-biodiversity-extinction-food-waste-health-meat-plant-based/a-56416006; and Damian Carrington, "Earth's Sixth Mass Extinction Event Under Way, Scientists Warn," *Guardian*, July 10, 2017, https://www.theguardian.com/environment/2017/jul/10/earths-sixth-mass-extinction-event-already-underway-scientists-warn.

14. United Nations, "Biodiversity Commitment Builds Hope for 'Living in Harmony with Nature,'" *UN News*, October 13, 2021, https://news.un.org/en/story/2021/10/1102942.

15. John R. Platt, "The Rise of Extinction Denial," *Scientific American*, June 22, 2019, https://blogs.scientificamerican.com/extinction-countdown/rise-of-the-extinction-deniers/.

16. Alexander C. Lees, "Biodiversity Scientists Must Fight the Creeping Rise of Extinction Denial," *Nature Ecology & Evolution* 4, no. 11 (2020): 1440–1443.

17. Roger Pielke Jr., "The Energy Transition Has Not Yet Started," *The Honest Broker*, June 29, 2023, https://rogerpielkejr.substack.com/p/the-energy-transition-has-not-yet; and IPSOS, *Climate Change: A Growing Skepticism*, December 9, 2022, https://www.ipsos.com/en-us/news-polls/obscop-2022.

18. Warren M. Herrn, *Homo Ecophagus: A Deep Diagnosis to Save the Earth* (New York: Routledge, 2023), 271.

19. E. O. Wilson, *Half-Earth: Our Planet's Fight for Life* (New York: Liveright, 2016), 3.

20. Arthur Koestler, *The Ghost in the Machine* (New York: Arkana, 1967), 267.

21. Kurt Vonnegut, *Galápagos* (New York: Dial Press, 2009), 296.

22. Vonnegut, *Galápagos*, 9.

23. Vonnegut, *Galápagos*, 278.

24. Vonnegut, *Galápagos*, 292.

25. Elizabeth Kolbert, *The Sixth Extinction: An Unnatural History* (New York: Henry Holt and Company, 2014), 251–252.

26. Paul J. Crutzen, "The 'Anthropocene,'" in *Earth System Science in the Anthropocene*, ed. Eckart Ehlers and Thomas Krafft (New York: Springer, 2006), 17.

27. Richardson et al., "Earth Beyond Six of Nine Planetary Boundaries."

28. Ben See (@ClimateBen), "OLD NEWS: Scientists say without an immediate shift away from capitalism global warming so hellish that it's beyond the ability of humans to adapt must be anticipated within 14 to 28 years," X, August 13, 2023, 2:42 p.m., https://x.com/ClimateBen/status/1690826135897010176.

29. William I. Boarman, *Threats to Desert Tortoise Populations: A Critical Review of the Literature* (Sacramento, CA: US Geological Survey, Western Ecological Research Center, 2002), 39.

30. Leo Hickman, "James Lovelock: Humans Are Too Stupid to Prevent Climate Change," *Guardian*, March 29, 2010, https://www.theguardian.com/science/2010/mar/29/james-lovelock-climate-change.

31. For instance, Ross Mittiga, "Political Legitimacy, Authoritarianism, and Climate Change," *Political Science Review* 116, no. 3 (2021): 1–14.

32. Cameron Abadl, "What if Democracy and Climate Mitigation Are Incompatible?," *Foreign Policy*, January 7, 2022, https://foreignpolicy.com/2022/01/07/climate-change-democracy/.

33. Arthur Nelsen, "We Need Power to Prescribe Climate Policy, IPCC Scientists Say," *Guardian*, December 7, 2023, https://www.theguardian.com/environment/2023/dec/07/we-need-power-to-prescribe-climate-policy-ipcc-scientists-say#:~:text=Five%20lead%20authors%20of%20IPCC,on%20climate%20change%20(UNFCCC).

34. Alex Trembath, "We Can't Sue Our Way to a Stable Climate," *Breakthrough Journal*, May 20, 2024, https://thebreakthroughjournal.substack.com/p/we-cant-sue-our-way-to-a-stable-climate?r=3ez4l&triedRedirect=true.

35. Christopher Ketcham, "How the Sage Grouse Became a Casualty of the Western Land War," *Westword*, July 23, 2019, https://www.westword.com/news/this-land-the-greater-sage-grouse-became-a-casualty-of-the-western-land-war-11415895.

36. Jeffrey Friedman, "Post-Truth and the Epistemological Crisis," *Critical Review* 35, no. 1–2 (2023): 1–21.

37. Naomi Oreskes and Erik M. Conway, *Merchants of Doubt: How a Handful of Scientists Obscured the Truth on Issues from Tobacco Smoke to Climate Change* (New York: Bloomsbury, 2011).

38. I have stopped assigning *Merchants of Doubt* to students largely because it, perhaps inadvertently, ends up pandering to their preferred storyline regarding why other people disagree with them. If I ever include it in the classroom again, it'll only be if I also include criticisms of the entire misinformation framework as well as evidence that Oreskes's own public writings make selective appeals to the facts, exaggerate certainties, and engage in concept creep to make disagreement seem illegitimate. Perhaps that those are lesser sins than those committed by doubt merchants, but it's not how I would like to see science used in a democracy. See David Ropenik, "Is Naomi Oreskes Using the Same Merchant of Doubt Tactics She Criticizes?," *Big Think*, December 19, 2015, https://bigthink.com/technology-innovation/is-naomi-oreskes-using-the-same-merchant-of-doubt-tactics-she-criticizes/; and Ted Nordhaus, "Demons Under Every Rock," *The Breakthrough Institute*, May 4, 2017, https://thebreakthrough.org/articles/demons-under-every-rock.

39. See Arthur Grundmann, "Debunking Sceptical Propaganda," *BioSocieties* 8, no. 3 (2013): 370–174.

40. Clare Carlile, "Sowing Doubt: How Big Ag is Delaying Sustainable Farming in Europe," *DeSmog*, December 21, 2022, https://www.desmog.com/2022/12/21/sowing-doubt-how-big-ag-is-delaying-sustainable-farming-in-europe/.

41. James Pogue, *Chosen Country: A Rebellion in the West* (New York: Henry Holt and Company, 2018), 78–95; the bigger issues for Nevada ranchers seem to have been that grazing fees didn't cover the area's management costs, see Peter S. Alagona, *After the Grizzly: Endangered Species and the Politics of Place in California* (Berkeley: University of California Press, 2013), 171.

42. Pogue, *Chosen Country*, 78–95; and Ketcham, *This Land.*

43. Scholarly readers should refer to my endnotes, like this one, for academic digressions and additional arguments that situate this book with regard to other works on environmental politics.

44. See Keith Kloor, "The Battle for the Soul of Conservation Science," *ISSUES in Science and Technology* 31, no. 1 (2015): 74–79.

45. Mark Vellend, "The Biodiversity Conservation Paradox," *American Scientist*, March–April (2017): 94–101.

46. For an accessible overview regarding why, see Kolbert, *The Sixth Extinction*, 148–153.

47. David Tilman, Forest Isbell, and Jane M. Cowles, "Biodiversity and Ecosystem Functioning," *Annual Review of Ecology, Evolution, and Systematics* 45, no. 1 (2014): 471–493.

48. Donald S. Maier, *What's So Good About Biodiversity? A Call for Better Reasoning About Nature* (New York: Springer, 2012), 349.

49. Maier, *What's So Good About Biodiversity*, 421–430. Maier calls this the "appropriate fit" theory of natural value, an approach that emphasizes the relationship between humans and nature.

50. Ranee Seither, Oyindamola Bidemi Yusuf, Devon Dramann, Kayla Calhoun, Agnes Mugerwa-Kasujja, and Cynthia L. Knighton, "Coverage with Selected Vaccines and Exemption from School Vaccine Requirements Among Children in Kindergarten—United States, 2022–23 School Year," *Morbidity and Mortality Weekly Report* 72 (2023): 1217–1224; and Craig Palosky, "More than 4 in 10 Republicans and a Third of Parents Now Oppose Schools Requiring Children to Get Vaccinated for Measles and Other Illness, Up Since the COVID-19 Pandemic Began," *KFF*, December 16, 2022, https://www.kff.org/coronavirus-covid-19/press-release/more-than-4-in-10-republicans-and-a-third-of-parents-now-oppose-schools-requiring-children-to-get-vaccinated-for-measles-and-other-illness-up-since-the-covid-19-pandemic-began/.

51. Brian Kennedy and Alex Tyson, "Americans' Trust in Scientists, Positive Views of Science Continue to Decline," *Pew Research Center*, November 14, 2023, https://www.pewresearch.org/science/2023/11/14/americans-trust-in-scientists-positive-views-of-science-continue-to-decline/.

52. As quoted in Sophia Rosenfeld, *Democracy and Truth* (Philadelphia: University of Pennsylvania Press), 13.

53. Robert P. Crease with Peter D. Bond, *The Leak: Politics, Activists, and Loss of Trust at Brookhaven National Laboratory* (Cambridge, MA: MIT Press, 2022).

54. Taylor Dotson, "Crisis Science," *Taming Complexity*, July 6, 2023, https://tamingcomplexity.substack.com/p/crisis-science.

55. Misinformation is more a problem of demand than one of supply. See Ari Schulman's comments in "What Comes Next in the Information Wars?," *American Enterprise Institute*, December 5, 2024, https://www.aei.org/events/what-comes-next-in-the-information-wars/.

CHAPTER 2

1. D. T. Max, "Green Is Good," *New Yorker*, May 15, 2014, https://www.newyorker.com/magazine/2014/05/12/green-is-good/.

2. As quoted in Max, "Green Is Good." See Michelle Marvier, Peter Kareiva, and Robert Lalasz, "Conservation in the Anthropocene," *The Breakthrough Journal* 2 (2012), https://thebreakthrough.org/journal/issue-2/conservation-in-the-anthropocene; and Tom Kunkel, "A Man of the People," *Nature Conservancy*, Spring (2011): 37.

3. As quoted in Max, "Green Is Good."

4. George Wuerthner, Eileen Crist, and Tom Butler, *Keeping the Wild: Against the Domestication of Earth* (Washington, DC: Island Press, 2014); and Stuart Pimm, "Keeping the Wild: Against the Domestication of Earth," *Biological Conservation* 180 (2014): 151–152.

5. Keith Kloor, "The Science Police," *ISSUES in Science and Technology* 33, no. 4 (2017), https://issues.org/the-science-police/.

6. Mark Vellend, Lander Baeten, Isla H. Myers-Smith, Sarah C. Elmendorf, Robin Beauséjour, Carissa D. Brown, et al., "Global Meta-Analysis Reveals No Net Change in Local-Scale Plant Biodiversity over Time," *Proceedings of the National Academy of Sciences* 110, no. 48 (2013): 19456–19459; and Maria Dornelas, Nicholas J. Gotelli, Brian McGill, Hideyasu Shimadzu, Faye Moyes, Caya Sievers, et al., "Assemblage Time Series Reveal Biodiversity Change but Not Systemic Loss," *Science* 344, no. 6181 (2014): 296–299.

7. Mark Vellend, "Is Biodiversity Declining on Local Scales?," in *Effective Conservation Science: Data Not Dogma*, ed. Peter Kareiva, Michelle Marvier, and Brian Silliman (New York: Oxford University Press, 2018), 30.

8. Andrew Gonzalez, Bradley J. Cardinale, Ginger R. H. Allington, Jarrett Byrnes, K. Arthur Endsley, Daniel G. Brown, et al., "Estimating Local Biodiversity Change: A Critique of Papers Claiming No Net Loss of Local Diversity," *Ecology* 97, no. 8 (2016): 1949–1960.

9. Mark Vellend, Maria Dornelas, Lander Baeten, Robin Beauséjour, Carissa D. Brown, Pieter De Frenne, et al., "Estimates of Local Biodiversity Change over Time Stand Up to Scrutiny," *Ecology* 98, no. 2 (2017): 583–590; Richard B. Primack, "Biodiversity Gains? The Debate

on Changes in Local- vs Global-Scale Species Richness," *Biological Conservation* 219 (2018): A1–A3; and Bradley J. Cardinale, Andrew Gonzalez, Ginger R. H. Allington, and Michel Loreau, "Is Local Biodiversity Declining or Not? A Summary of the Debate over Analysis of Species Richness Time Trends," *Biological Conservation* 219 (2018): 175–183.

10. Jose W. Valdez, Corey T. Callaghan, Jessica Junker, Andy Purvis, Samantha L. L. Hill, Henrique M. Pereira, "The Undetectability of Global Biodiversity Trends Using Local Species Richness," *Ecography* 2023, no. 3 (2023): e06604, https://doi.org/10.1111/ecog.06604.

11. Dov F. Sax and Steven D. Gaines, "Species Diversity: From Global Decreases to Local Increases," *Trends in Ecology & Evolution* 18, no. 11 (2003): 561–566; and Helmut Hillebrand, Bernd Blasius, Elizabeth T. Borer, Jonathan M. Chase, John A. Downing, Britas Klemens Eriksson, et al., "Biodiversity Change Is Uncoupled from Species Richness Trends: Consequences for Conservation and Monitoring," *Journal of Applied Ecology* 55, no. 1 (2018): 169–184.

12. R. Alexander Pyron, "We Don't Need to Save Endangered Species. Extinction Is Part of Evolution," *Washington Post*, November 22, 2017, https://www.washingtonpost.com/outlook/we-dont-need-to-save-endangered-species-extinction-is-part-of-evolution/2017/11/21/57fc5658-cdb4-11e7-a1a3-0d1e45a6de3d_story.html.

13. A study published as I was finishing this book found that instances of biological homogenization approximately balanced cases of biological differentiation. The authors recommend "moving beyond a belief of ubiquitous biodiversity loss and homogenization." Shane A. Blowes, Brian McGill, Viviana Brambilla, Cher F. Y. Chow, Thore Engel, Ada Fontrodona-Eslava, et al., "Synthesis Reveals Approximately Balanced Biotic Differentiation and Homogenization," *Science Advances* 10, no. 8 (2024), https://doi.org/10.1126/sciadv.adj9395.

14. Barnabas H. Daru, T. Jonathan Davies, Charles G. Willis, Emily K. Meineke, Argo Ronk, Martin Zobel, et al., "Widespread Homogenization of Plant Communities in the Anthropocene," *Nature Communications* 12, no. 6983 (2021), https://doi.org/10.1038/s41467-021-27186-8; and Michel Loreau, Nicolas Mouquet, and Andrew Gonzalez, "Biodiversity as Spatial Insurance in Heterogeneous Landscapes," *Proceedings of the National Academy of Sciences* 100, no. 22 (2003): 12765–12770.

15. Tim Newbold, Lawrence N. Hudson, Samantha L. L. Hill, Sara Contu, Igor Lysenko, Rebecca A. Senior, et al., "Global Effects of Land Use on Local Terrestrial Biodiversity," *Nature* 520, no. 7545 (2015): 45–50.

16. Steve Rayner and Daniel Sarewitz, "Policy Making in the Post-Truth World: On the Limits of Science and the Rise of Inappropriate Expertise," *The Breakthrough Journal* Winter (2021): 15–43; and Daniel Sarewitz, "How Science Makes Environmental Controversies Worse," *Environmental Science & Policy* 7, no. 5 (2004): 385–403.

17. Primack, "Biodiversity Gains?," A2.

18. Elizabeth Pop Berman, *Thinking like an Economist: How Efficiency Replaced Equality in U.S. Public Policy* (Princeton, NJ: Princeton University Press, 2022), 46–48.

19. *Pandora's Box: A Fable from the Age of Science*, episode 2, "To the Brink of Eternity," written by Adam Curtis, aired June 18, 1992, YouTube, 2017, https://www.youtube.com/watch?v=34T8s0D-Fxo&t=3s; and Ben Connable, *Embracing the Fog of War* (Santa Monica, CA: RAND, 2012).

20. Alexis C. Madrigal, "The Computer that Predicted the U.S. Would Win the Vietnam War," *Atlantic*, October 5, 2017, https://www.theatlantic.com/technology/archive/2017/10/the-computer-that-predicted-the-us-would-win-the-vietnam-war/542046/.

21. David R. Jardini, "Out of the Blue Yonder: The RAND Corporation's Diversification into Social Welfare Research 1946–1968" (PhD diss., Carnegie Mellon University, 1996), 108 (my italics).

22. I borrow the phrase from Brad East, who uses it in a review of Jason Blakely's analysis of how the social sciences has "'infiltrated'" our common life of politics and ethics." Brad East, "Statistics as Storytelling," *New Atlantis* 66 (2021): 111–122.

23. Damian Carrington, "Humanity Has Wiped Out 60% of Animal Populations Since 1970, Report Finds," *Guardian*, October 29, 2018, https://www.theguardian.com/environment/2018/oct/30/humanity-wiped-out-animals-since-1970-major-report-finds.

24. Brian Leung, Anna L. Hargreaves, Dan A. Greenberg, Brian McGill, Maria Dornelas, and Robin Freeman, "Clustered Versus Catastrophic Global Vertebrate Declines," *Nature* 588 (2020): 267–271; and Hannah Ritchie and Max Roser, "Living Planet Index," *Our World in Data*, https://ourworldindata.org/living-planet-index.

25. For the LPI, the mean is calculated on rates of population changes, and the logarithmic form of the calculation is used.

26. Author creation. Data sourced from Leung et al., "Clustered Versus Catastrophic."

27. Leung et al., "Clustered Versus Catastrophic," 267–271.

28. Cf. Gopal Murali, Gabriel Henrique de Oliveira Caetano, Goni Barki, Shai Meiri, and Uri Roll, "Emphasizing Declining Populations in the Living Planet Report," *Nature* 601 (2022): E20–E24; Zia Mehrabi and Robin Naidoo, "Shifting Baselines and Biodiversity Success Stories," *Nature* 601 (2022): E17–E18; and Michel Loreau, Bradley J. Cardinale, Forest Isbell, Tim Newbold, Mary I. O'Connor, and Claire de Mazancourt, "Do Not Downplay Biodiversity Loss," *Nature* 601 (2022): E27–E28.

29. Loreau et al., "Do Not Downplay," E28.

30. Mehrabi and Naidoo, "Shifting Baselines," E18.

31. I've focused on cases where pessimism is baked into the model. However, the opposite is sometimes the case. See Orrin H. Pilkey and Linda Pilkey-Jarvis, *Useless Arithmetic: Why Environmental Scientists Can't Predict the Future* (New York: Columbia University Press, 2007).

32. Rayner and Sarewitz, "Policy Making in the Post-Truth World," 28.

33. David L. Wagner, Eliza M. Grames, Matthew L. Forister, May R. Berenbaum, and David Stopak, "Insect Decline in the Anthropocene: Death by a Thousand Cuts," *Proceedings of the National Academy of Sciences* 118, no. 2 (2021): e2023989118.

34. Caspar A. Hallmann, Martin Sorg, Eelke Jongejans, Henk Siepel, Nick Hofland, Heinz Schwan, et al., "More than 75 Percent Decline over 27 Years in Total Flying Insect Biomass in Protected Areas," *PLoS One* 12, no. 10 (2017): e0185809, https://doi.org/10.1371/journal.pone.0185809.

35. Francisco Sánchez-Bayo and Kris A. G. Wyckhuys, "Worldwide Decline of the Entomofauna: A Review of its Drivers," *Biological Conservation* 232 (2019): 8–27.

36. Brooke Jarvis, "The Insect Apocalypse Is Here," *New York Times*, November 27, 2018, https://www.nytimes.com/2018/11/27/magazine/insect-apocalypse.html; George Monbiot, "Insectageddon: Farming Is More Catastrophic than Climate Breakdown," *Guardian*, October 20, 2017, https://www.theguardian.com/commentisfree/2017/oct/20/insectageddon-farming-catastrophe-climate-breakdown-insect-populations; and Florian Cazeres, "German Farmers Rail Against Insect Protection Plans," *Phys.org*, February 10, 2021, https://phys.org/news/2021-02-german-farmers-rail-insect.html.

37. Silke Beck, "Moving Beyond the Linear Model of Expertise? IPCC and the Test of Adaptation," *Regional Environmental Change* 11 (2011): 297–306.

38. Sarewitz, "How Science Makes Environmental Controversies Worse," 388.

39. Taylor Dotson and Michael Bouchey, "Democracy and the Nuclear Stalemate," *New Atlantis*, Fall (2020): 14–45.

40. See, e.g., Vincent Ni, "'Tit for Tat': Why Hunt for Covid's Origins Still Mired in Politics and Controversy," *Guardian*, December 31, 2021, https://www.theguardian.com/world/2021/dec/31/why-hunt-for-covid-origins-still-wrapped-in-politics-impasse-china-west.

41. Taylor Dotson, *The Divide: How Fanatical Certitude Is Destroying Democracy* (Cambridge, MA: MIT Press, 2021).

42. Geert De Clercq, "European Parliament Scraps Pesticides Bill in Latest Setback for Green Legislation," *Reuters*, November 22, 2023, https://www.reuters.com/world/europe/european-parliament-scraps-pesticides-bill-latest-setback-green-legislation-2023-11-22/#:~:text=With%20299%20votes%20against%20the,been%20negotiated%20intensely%20for%20months.

43. Corporate Europe Observatory, "A Loud Lobby for a Silent Spring," March 17, 2022, https://corporateeurope.org/en/2022/03/loud-lobby-silent-spring. This presents an obviously partisan view of the dispute, but the "follow the science" dynamic is still clear.

44. Lees et al., "Biodiversity Scientists Must Fight."

45. James C. Russell and Tim M. Blackburn, "The Rise of Invasive Species Denialism," *Trends in Ecology & Evolution* 32, no. 1 (2017): 3–6.

46. Ross N. Cuthbert, Sven Bacher, Tim M. Blackburn, Elizabeta Briski, Christophe Diagne, Jaimie T. A. Dick, et al., "Invasion Costs, Impacts, and Human Agency: Response to Sagoff 2020," *Conservation Biology* 34, no. 6 (2020): 1579–1582.

47. Matthew C. Nisbet, "Sciences, Publics, Politics: The Trouble with Climate Emergency Journalism," *ISSUES in Science and Technology* 35, no. 4 (2019): 23–26; and Kloor, "The Science Police."

48. Matthew G. Burgess, Roger Pielke Jr., and Justin Ritchie, "Catastrophic Climate Risks Should Be Neither Understated nor Overstated," *Proceedings of the National Academy of Sciences* 119, no. 42 (2022): e2214347119; and Roger Pielke Jr., "Opening Up the Climate Policy Envelope," *ISSUES in Science and Technology* 34, no. 4 (2018), https://issues.org/opening-up-the-climate-policy-envelope/. For an example of a paper predicting biodiversity collapse under climate change using RCP 8.5, see Alaaeldin Soultan, Martin Wikelski, and Kamran Safi, "Risk of Biodiversity Collapse Under Climate Change in the Afro-Arabian Region," *Scientific Reports* 9, no. 955 (2019), https://doi.org/10.1038/s41598-018-37851-6.

49. Reiner Grundmann, "The Legacy of Climategate," *WIREs Climate Change* 3 (2012): 281–288; and Justin Gillis, "British Panel Clears Scientists," *New York Times*, July 7, 2010, https://www.nytimes.com/2010/07/08/science/earth/08climate.html.

50. Michael S. Crossley, Amanda R. Meier, Emily M. Baldwin, Lauren L. Berry, Leah C. Crenshaw, Glen L. Hartman, et al., "No Net Insect Abundance and Diversity Declines Across US Long Term Ecological Research Sites," *Nature Ecology & Evolution* 4 (2020): 1368–1376; and Roel van Klink, Diana E. Bowler, Konstantin B. Gongalsky, Ann B. Swengel, Alessandro Gentile, and Jonathan M. Chase, "Meta-Analysis Reveals Declines in Terrestrial but Increases in Freshwater Insect Abundances," *Science* 368, no. 6489 (2020): 417–420.

51. David L. Wagner, "Global Insect Decline: Comments on Sánchez-Bayo and Wyckhuys (2019)," *Biological Conservation* 233 (2019): 332–333; and Manu E. Saunders, Jasmine K. Janes, and James C O'Hanlon, "Moving On from the Insect Apocalypse Narrative," *BioScience* 70, no. 1 (2020): 80–89.

52. Diana E. Bowler, David Eichenberg, Klaus-Jürgen Conze, Frank Suhling, Kathrin Baumann, Theodor Benken, et al., "Winners and Losers over 35 Years of Dragonfly and Damselfly Distributional Change in Germany," *Diversity and Distributions* 27, no. 8 (2021): 1353–1366.

53. Manu E. Saunders, "No Simple Answers for Insect Conservation," *American Scientist* 107, no. 3 (2019): 148.

54. Unsurprisingly, the German government backtracked when the full ban was due to come into effect. Madeline Chambers, "German Cabinet Approves Restricted Use of Herbicide Glyphosate," *Reuters*, April 24, 2024, https://www.reuters.com/world/europe/german-cabinet-approves-restricted-use-herbicide-glyphosate-2024-04-24/.

55. Caspar A. Hallmann, Theo Zeegers, Roel van Klink, Rikjan Vermeulen, Paul van Wielink, Henk Spijkers, et al., "Declining Abundance of Beetles, Moths and Caddisflies in the Netherlands," *Insect Conservation and Diversity* 13, no. 2 (2020): 127–139.

56. Juliet Eilperin and Steven Mufson, "Keystone XL Pipeline Would Have Little Impact on Climate Change, State Dept. Says," *Washington Post*, March 1, 2013, https://www.washingtonpost.com/national/health-science/keystone-xl-pipeline-will-not-have-huge-impact-on-climate-draft-analysis-says/2013/03/01/715491b0-82a5-11e2-b99e-6baf4ebe42df_story.html.

57. Jörg Müller, "Weather Explains the Decline and Rise of Insect Biomass over 34 Years," *Nature* 553 (2023), https://doi.org/10.1038/s41586-023-06402-z. Note: Evidence of insect biomass increases came from different research sites than those used by Hallman et al., making direct comparisons difficult.

58. Heinz Horeis, "Once upon a Time, There Was a Dying Forest," *Technology in Society* 31, no. 1 (2009): 111–116.

59. *More of Less: Behind the Stats*, "Insectageddon," produced by Darin Graham, presented by Ruth Alexander, released on March 2, 2019, BBC Radio, podcast, https://www.bbc.co.uk/sounds/play/p072c44x/.

60. And they're not the only ones. Citizens often yearn for authoritative science to wash away discomforting uncertainties. See Senja Post, Nils Bienzeisler, and Mareike Lohöfener, "A Desire for Authoritative Science? How Citizens' Informational Needs and Epistemic Beliefs Shaped Their Views of Science, News, and Policymaking in the COVID-19 Pandemic," *Public Understanding of Science* 30, no. 5 (2021): 496–514.

61. Hannah Ritchie, "There Have Been Five Mass Extinctions in Earth's History," November 30, 2020, https://ourworldindata.org/mass-extinctions, used under Creative Commons Attributional 4.0 International license, https://creativecommons.org/licenses/by/4.0/https://creativecommons.org/licenses/by-sa/4.0/deed.en. Modifications by author: Data extracted, regraphed in Excel, and depicted in grayscale without listing all geological eras. Original data sources: Anthony D. Barnosky, Nicholas Matzke, Susumu Tomiya, Guinevere O. U. Wogan, Brian Swartz, Tiago B. Quental, et al., "Has the Earth's Sixth Mass Extinction Already Arrived?," *Nature* 471, no. 7336 (2011): 51–57; and Malcolm M. McCallum, "Vertebrate Biodiversity Losses Point to a Sixth Mass Extinction," *Biodiversity and Conservation* 24, no. 10 (2015): 2497–2519.

62. Lee Sweetlove, "Number of Species on Earth Tagged at 8.7 Million," *Nature* (2011), https://doi.org/10.1038/news.2011.498; David Sephokoski, *Catastrophic Thinking* (Chicago: University of Chicago Press, 2020), 229–294; Katarina Zimmer, "Are We in the Midst of a Sixth Mass Extinction?," *Scientist*, July 18, 2022, https://www.the-scientist.com/features/are-we-in-the-midst-of-a-sixth-mass-extinction-70172; and Valentí Rull, "Biodiversity Crisis or Sixth Mass Extinction?," *EMBO Reports* 23, no. 1 (2022): e54193.

63. Zimmer, "Are We in the Midst"; and Rull, "Biodiversity Crisis."

64. Robert H. Cowie, Philippe Bouchet, and Benoît Fontaine, "The Sixth Mass Extinction: Fact, Fiction or Speculation?," *Biological Reviews* 29 (2022): 640–663; and John C. Briggs, "Emergence of a Sixth Mass Extinction?," *Biological Journal of the Linnean Society* 122, no. 2 (2017): 243–248.

65. The divide between Cowie's group and John Briggs isn't just a matter of using different methods to estimate how many species are dying out. They have completely different positions regarding the politics of biodiversity. Cowie's paper defends against charges of overestimation, declaring that their projections only seem so "to those who reject facts that do not conform to their personal or political agendas." And they accuse biologists such as Chris Thomas of "[giving] carte blanche to those who would destroy the Earth for their own short-term game"; Cowie et al., "The Sixth Mass Extinction," 655. Briggs contends that sixth-extinction talk detracts from the more pressing problem of protecting habitats for specific threatened species, arguing that "misleading the public [with exaggerated estimates] is not an effective conservation strategy"; John C. Briggs, "Global Biodiversity Loss: Exaggerated Versus Realistic Estimates," *Environmental Skeptics and Critics* 5, no. 2 (2016): 25.

66. Fred Pearce, "Global Extinction Rates: Why Do Estimates Vary So Wildly?," *YaleEnvironment360*, August 17, 2015, https://e360.yale.edu/features/global_extinction_rates_why_do_estimates_vary_so_wildly.

67. Barnosky et al., "Has the Earth's Sixth Mass Extinction Already Arrived?"

68. David Adam, "How Far Will Global Population Rise? Researchers Can't Agree," *Nature* 597, no. 7877 (2021): 462–465.

69. As quoted in Louis Bergeron, "Discovering Mammals Cause for Worry," *Stanford News*, February 9, 2009, https://news.stanford.edu/news/2009/february11/numa-021109.html.

70. Or so argues Barry W. Brook, Erle C. Ellis, and Jessie C. Buettel, "What Is the Evidence for Planetary Tipping Points?," in *Effective Conservation Science: Data Not Dogma*, ed. Peter Kareiva, Michelle Marvier, and Brian Silliman (New York: Oxford University Press, 2018), 51–57.

71. Scott Dance, "Maui Fires Not Just Due to Climate Change but a 'Compound Disaster,'" *Washington Post*, August 12, 2023, https://www.washingtonpost.com/weather/2023/08/12/hawaii-fires-climate-change-maui/; and Brianna Sacks, "Power Lines Likely Caused Maui's First Reported Fire, Video and Data Show," *Washington Post*, August 15, 2023, https://www.washingtonpost.com/climate-environment/2023/08/15/maui-fires-power-line-cause/. See also Fred Pearce, "It's Not Just Climate: Are We Ignoring Other Causes of Disasters?," *YaleEnvironment360*, February 8, 2022, https://e360.yale.edu/features/its-not-just-climate-are-we-ignoring-other-causes-of-disasters.

72. Sarah Bahr, "Tracking Climate Change in 193 Countries," *New York Times*, January 2, 2022, https://www.nytimes.com/2022/01/02/insider/tracking-climate-change-in-193-countries.html?partner=IFTTT.

73. I made a similar argument in the context of COVID-19 in Taylor Dotson, "Radiation Politics in a Pandemic," *New Atlantis*, June 18, 2020, https://www.thenewatlantis.com/publications/radiation-politics-in-a-pandemic/.

74. As quoted in Michael Hannon, "The Politics of Post-Truth," *Critical Review* 25, no. 1–2 (2023): 53.

75. Steve Hanley, "Hansen vs. Mann—Is Global Warming Linear or Exponential," *CleanTechnica*, November 3, 2023, https://cleantechnica.com/2023/11/03/hansen-vs-mann-is-global-warming-linear-or-exponential/. Worries that the Atlantic Meridional Ocean Current was already collapsing were based on sea surface temperatures, but an analysis based on air-sea heat fluxes finds the current to be far more stable than previously believed. Jens Tehaar, Linus Vogt, and Nicholas P. Foukal, "Atlantic Overturning Inferred from Air-Sea Heat Fluxes Indicates No Decline Since the 1960s," *Nature Communications* 16, no. 222 (2025), https://doi.org/10.1038/s41467-024-55297-5. As Roger Pielke Jr. has pointed out, the creation and loss of coastline are complex processes, with the global trend actually tending toward a gain of shoreline, even for atoll islands. Roger Pielke Jr., "Landification," *The Honest Broker*, January 15, 2024, https://rogerpielkejr.substack.com/p/landification. For more on the challenges of expert knowledge, see Jeffrey Friedman, *Power Without Knowledge: A Critique of Technocracy* (New York: Oxford University Press, 2020).

76. Ted Nordhaus, "Did Exxon Make It Rain Today?," *New Atlantis*, Winter 2024, https://www.thenewatlantis.com/publications/did-exxon-make-it-rain-today. Moreover, we should remember that an additional cause of ballooning disaster costs is the earth's growing human population and the tendency to build evermore settlements along rivers and coastlines.

77. As quoted in *Guardian*, "'I Want You to Panic': 16-Year-Old Issues Climate Warning at Davos," January 25, 2019, https://www.theguardian.com/science/video/2019/jan/25/i-want-you-to-panic-16-year-old-greta-thunberg-issues-climate-warning-at-davos-video.

78. Roger Pielke Jr., *The Honest Broker: Making Sense of Science in Policy and Politics* (New York: Cambridge University Press, 2007), 40–50.

79. Nicholas Harris, "Greta Thunberg Throws in Her Lot with the Anti-Capitalist Left," *UnHerd*, October 31, 2022, https://unherd.com/thepost/greta-thunberg-throws-her-lot-in-with-the-anti-capitalist-left/.

80. Alexander Zaitchik, "The Urgent Case for Shrinking the Economy," *The New Republic*, December 28, 2020, https://newrepublic.com/article/160692/less-is-more-degrowth-climate-change-book-review; and Richard Smith, "Climate Crisis, the Deindustrialization Imperative and the Jobs vs. Environment Dilemma," *Truthout*, November 12, 2014, https://truthout.org/articles/climate-crisis-the-deindustrialization-imperative-and-the-jobs-vs-environment-dilemma/.

81. See, e.g., Ellen Goodman, "Breaking Down Political, Psychological Barriers to Global Warming Action," *Baltimore Sun*, February 8, 2007, https://www.baltimoresun.com/news/bs-xpm-2007-02-09-0702090234-story.html/.

82. Pielke, *The Honest Broker*, 40–50.

83. Mike Hulme, *Why We Disagree About Climate Change* (Cambridge: Cambridge University Press, 2009).

84. See Alexander Bogner, *Die Epistemisierung des Politischen: Wie die Macht des Wissens die Demokratie Gefährdet* (Ditzingen, Germany: Philipp Reclam, 2021), 19–21.

85. Taylor Dotson, "Unsustainable Alarmism," *New Atlantis* 69 (2022): 41–48.

86. Veegallan (@AllanVMurray), Twitter (now X), April 4, 2022, https://twitter.com/AllanVMurray/status/1511013243199836162?ref_src=twsrc%5Etfw.

87. Sven Hillenkamp, "Die Panik vor den Rettern, " *Zeit Online*, March 25, 2022, https://www.zeit.de%2Fkultur%2F2022-03%2Ffridays-for-future-klimastreik-klimabewegung-klimawandel/.

88. Cameron Abadi, "What If Democracy and Climate Mitigation Are Incompatible," *Foreign Policy*, January 7, 2022; and Magdalena Hirsch, "Becoming Authoritarian for the Greater Good? Authoritarian Attitudes in Context of the Societal Crises of COVID-19 and Climate Change," *Frontiers in Political Science* 23 (2022), https://doi.org/10.3389/fpos.2022.929991.

89. Dotson, "Unsustainable Alarmism."

90. Joseph Poore, "How Can We Cut Soaring Demand for Meat? Try a Hybrid Burger," *Guardian*, November 7, 2022, https://www.theguardian.com/commentisfree/2022/nov/07/soaring-demand-meat-hybrid-burger-blending-benefits.

91. Sephokoski, *Catastrophic Thinking*, 263.

92. Ted Nordhaus, "Twilight of Environmental Idols," *The Breakthrough Institute*, September 26, 2022, https://thebreakthrough.org/issues/energy/twilight-environmental-idols-what-comes-after-environmentalism.

93. Michael E. Soulé, "What Is Conservation Biology?," *BioScience* 35, no. 11 (1985): 727–734.

94. David Takacs, *The Idea of Biodiversity* (Baltimore: Johns Hopkins University Press, 1996).

95. Takacs, *The Idea of Biodiversity*, 250.

96. Takacs, *The Idea of Biodiversity*, 259.

97. As quoted in Emma Marris, *Rambunctious Garden* (New York: Bloomsbury, 2011), 166.

98. Takacs, *The Idea of Biodiversity*, 152–154.

99. Takacs, *The Idea of Biodiversity*, 96.

100. As quoted in Frederic C. Rich, *Getting to Green* (New York: W. W. Norton, 2016), 165.

101. Takacs, *The Idea of Biodiversity*, 123–133. See also Sephokoski, *Catastrophic Thinking*.

102. Takacs, *The Idea of Biodiversity*, 2.

103. Sarewitz, "How Science Makes Environmental Controversies Worse," 392.

104. Ted Nordhaus and Michael Shellenberger, *Break Through: From the Death of Environmentalism to the Politics of Possibility* (New York: Houghton Mifflin, 2007), 132.

105. Kevin Elliott, *A Tapestry of Values: An Introduction to Values in Science* (New York: Oxford University Press, 2017).

106. Abraham J. Miller-Rushing, Richard B. Primack, Vincent Devictor, Richard T. Corlett, Graeme S. Cumming, Rafael Loyola, et al., "How Does Habitat Fragmentation Affect Biodiversity?," *Biological Conservation* 232 (2019): 271–273.

107. As Ester Turnhout, Willemijn Tuinstra, and Willem Halffman note, this "does not mean dismissing environmental knowledge" but rather "the recognition that nature and environment can be known and represented in diverse, but potentially equally legitimate, ways, none of which are neutral and value free." No doubt. I share their position that we should be wary of claims that one group of scientists' view of nature can be the Truth, but I think that science studies is often overly focused on making everything into a knowledge dispute. Ultimately, the conflict over nature is about competing ways of living and thus amounts to far more than merely the divergent environmental knowledge claims of different partisan groups. Any vision of nature is also one of world-building, not just world-seeing. And, as I point out later, I believe that democratic principles give local people the right to influence how their environments are managed, regardless of whether they possess relevant "lay expertise" for the situation. See Ester Turnhout, Willemijn Tuinstra, and Willem Halffman, *Environmental Expertise: Connecting Science, Policy, and Society* (New York: Cambridge University Press, 2019), 253.

108. Mark Vellend, "The Behavioral Economics of Biodiversity Conservation Scientists," *Philosophical Topics* 47, no. 1 (2019): 219–237.

109. Peter Kareiva and Michelle Marvier, "What Is Conservation Science?," *BioScience* 62, no. 11 (2012): 962–969.

110. See Kloor, "The Battle for the Soul."

111. Takacs, *The Idea of Biodiversity*, 328.

112. Takacs, *The Idea of Biodiversity*, 337.

113. Sephokoski, *Catastrophic Thinking*, 265.

114. A similar argument is made by Unai Pascual, William M. Adams, Sandra Díaz, Sharachchandra Lele, Georgina M. Mace, and Esther Turnhout, "Biodiversity and the Challenge of Pluralism," *Nature Sustainability* 4, no. 7 (2021): 567–572.

115. Dotson, "Unsustainable Alarmism"; and Dotson, *The Divide*.

116. Mike Hulme, "One Earth, Many Futures, No Destination," *One Earth* 2, no. 4 (2020): 309–311.

CHAPTER 3

1. Juan Pablo Escobar, *Pablo Escobar: My Father* (New York: MacMillan, 2016) is used throughout as my reference for biographical information on Pablo Escobar.

2. "HIPPOPOTAMI" by FICG.mx, *flickr*, March 16, 2011, https://flickr.com/photos/guadalajaracinemafest/5533572075/in/photolist-eXTQPx-eXTQ5X-eY6evq-viByW-qKRVNa-CQpsG5-9qZ1c6-2ojVCHN-2ojYmKU, used under Creative Commons Attribution 2.0 Generic license, https://creativecommons.org/licenses/by/2.0/.

3. Fernando Duarte, "Pablo Escobar: Why Scientists Want to Kill Colombia's Hippos," *BBC News*, February 11, 2021, https://www.bbc.com/news/world-latin-america-56011594.

4. Chris Kraul, "Killing of Hippo in Colombia Sparks Outcry," *Los Angeles Times*, July 15, 2009, https://www.latimes.com/archives/la-xpm-2009-jul-15-fg-colombia-hippos15-story.html.

5. Sebastían Restrepo and Carlos Daniel Cadena, "Science Denialism Limits Management of Invasive Hippos in Colombia," *Frontiers in Ecology and the Environment* 19, no. 6 (2021): 323. See also D. Nataly Castelblanco-Martínez, R. A. Moreno-Arias, J. A. Velasco, J. W. Moreno-Bernal, S. Restrepo, E. A. Noguera-Urbano, et al., "A Hippo in the Room: Predicting the Persistence and Dispersion of an Invasive Mega-Vertebrate in Colombia, South America," *Biological Conservation* 253 (2021): 108923.

6. Jessica L. Metcalf, Chris Turney, Ross Barnett, Fabiana Martin, Sarah C. Bray, Julia T. Vilstrup, et al., "Synergistic Roles of Climate Warming and Human Occupation in Patagonian Megafaunal Extinctions During the Last Deglaciation," *Science Advances* 2, no. 6 (2016), https://doi.org/10.1126/sciadv.1501682.

7. Anna J. Turbelin, Bruce D. Malamud, and Robert A. Francis, "Mapping the Global State of Invasive Alien Species: Patterns of Invasion and Policy Responses," *Global Ecology and Biogeography* 26, no. 1 (2017): 78–92.

8. *Cane Toads: An Unnatural History*, directed by Mark Lewis (1988; Australia, Film Australia), VHS.

9. Richard Shine and John J. Wiens, "The Ecological Impact of Invasive Cane Toads (Bufo Marinus) in Australia," *Quarterly Review of Biology* 85, no. 3 (2010): 253–291.

10. Thomas Lyttle, "Misuse and Legend in the 'Toad Licking' Phenomenon," *International Journal of the Addictions* 28, no. 6 (1993): 521–538.

11. Richard P. Duncan, "Magnitude and Variation of Prehistoric Bird Extinctions in the Pacific," *Proceedings of the National Academy of Sciences* 110, no. 16 (2013): 6436–6441.

12. Petr Pyšek, Philip E. Hulme, Dan Simberloff, Sven Bacher, Tim M. Blackburn, James T. Carlton, et al., "Scientists' Warning on Invasive Alien Species," *Biological Reviews* 95, no. 6 (2020): 1511–1534.

13. Christophe Diagne, Boris Leroy, Anne-Charlotte Vaissière, Rodolphe E. Gozlan, David Roiz, Ivan Jarić, et al., "High and Rising Economic Costs of Biological Invasions Worldwide," *Nature* 592, no. 7855 (2021): 571–576.

14. Patrick Greenfield, "Increase in Invasive Species Poses Dramatic Threat to Biodiversity—Report," *Guardian*, July 15, 2020, https://www.theguardian.com/environment/2020/jul/15/increase-in-invasive-species-poses-dramatic-threat-to-biodiversity-report-aoe.

15. Emma Marris, *Wild Souls* (New York: Bloomsbury, 2021), 192–193; and Elizabeth Kolbert, "The Big Kill," *New Yorker*, December 22, 2014, http://www.newyorker.com/magazine/2014/12/22/big-kill.

16. Roger L. Kitching, "A World of Thought: 'The Ecology of Invasions by Animals and Plants' and Charles Elton's Life's Work," in *Fifty Years of Invasion Ecology: The Legacy of Charles Elton*, ed. David M. Richardson (Hoboken, NJ: Blackwell, 2011), 5.

17. Charles S. Elton, *The Ecology of Invasions by Animals and Plants* (1958; repr., Chicago: University of Chicago Press, 2000).

18. Brendon M. H. Larson, "The War of the Roses: Demilitarizing Invasion Biology," *Frontiers in Ecology and the Environment* 3, no. 9 (2005): 496.

19. Martin A. Schlaepfer, "Do Non-Native Species Contribute to Biodiversity," *PLoS Biology* 16, no. 4 (2018): e2005568.

20. Adam Curtis, dir., "Goodbye Mrs. Ant," episode 4 of *Pandora's Box*, BBC Two, 1992.

21. Curtis, "Goodbye Mrs. Ant."

22. Radu Cornel Guiaşu, *Non-Native Species and Their Role in the Environment: The Need for a Broader Perspective* (Boston, MA: Brill, 2016), 203–243; Scott Palmer, "What Is 1080, and Why Do People Oppose It?," *Newshub*, August 30, 2022, https://www.newshub.co.nz/home/new-zealand/2018/08/what-is-1080-and-why-do-people-oppose-it.html; Kaushik Patowary, "Project Isabela," *Amusing Planet*, August 15, 2019, https://www.amusingplanet.com/2019/08/project-isabela-how-goats-helped.html; and Marris, *Rambunctious Garden*, 1–17.

23. Marris, *Wild Souls*, 192; and Marris, *Rambunctious Garden*, 11.

24. Ken Thompson, *Where Do Camels Belong?* (Vancouver, BC: Greystone, 2014); Sarah Bell, "Australia, Home to the World's Largest Camel Herd," *Guardian*, May 19, 2013, https://www.bbc.com/news/magazine-22522695; and Chris Heller, "Whatever Happened to the Wild Camels of the American West?," *Smithsonian Magazine*, August 6, 2015, https://www.smithsonianmag.com/history/whatever-happened-wild-camels-american-west-180956176/.

25. Miguel Claro, "Shifting Baselines and the Conservation of Non-Native Species," *Conservation Biology* 28, no. 5 (2014): 1434–1436; and Chris Thomas *Inheritors of the Earth, How Nature Is Thriving in an Age of Extinction* (New York: Public Affairs, 2017), 52.

26. Thomas, *Inheritors of the Earth*, 80–82.

27. Thomas, *Inheritors of the Earth*, 134–136.

28. Matthew Brown, "U.S. Officials Plan to Kill Hundreds of Thousands of Barred Owls to Save Another Species from Extinction," *PBS Newshour*, July 3, 2024, https://www.pbs.org/newshour/nation/u-s-officials-plan-to-kill-hundreds-of-thousands-of-barred-owls-to-save-another-species-from-extinction; Jeff Barnard, "Barred Owl Inroads, Hybrids Complicate Spotted Owl Review," *The Spokesman-Review*, June 20, 2004; Marris, *Wild Souls*, 183; and Christopher J. Preston, "Professional 'Owl Terrorists' Scare Off Barred Owls with Shotguns in the Name of Conservation," *Salon*, February 26, 2023, https://www.salon.com/2023/02/26/professional-owl-terrorists-scare-off-barred-owls-with-shotguns-in-the-name-of-conservation/.

29. Nicola Jones, "Pregnancy Test Helped to Bring Frog-Killing Fungus to The US," *Nature* (2013), https://doi.org/10.1038/nature.2013.13013.

30. Menno Schilthuizen, Lúcia P. Santos Pimenta, Youri Lammers, Peter J. Steenbergen, Marco Flohil, Nils G. P. Beveridge, et al., "Incorporation of an Invasive Plant into a Native Insect Herbivore Food Web," *PeerJ* 4 (2016): e1954.

31. Patrícia Rodrigues Brauun and Ana Maria Ribeiro, "Evolução dos Toxodontia da América do Sul Durante o Cenozoico: Aspectos Dentários, Paleoclimáticos e Paleoambientais," *Terrae Didactica* 13, no. 2 (2017): 131, used under Creative Commons Attributional 4.0 International license, https://creativecommons.org/licenses/by/4.0/.

32. Erick J. Lundgren, Daniel Ramp, John Rowan, Owen Middleton, Simon D. Schowanek, Oscar Sanisidro, et al., "Introduced Herbivores Restore Late Pleistocene Ecological Functions," *Proceedings of the National Academy of Sciences* 117, no. 14 (2020): 7871–7878.

33. Haley Cohen Gilliland, "Argentina Brought Beavers to Tierra Del Fuego. It Was Not a Good Idea," *National Geographic*, July 26, 2019, https://www.nationalgeographic.com/animals/article/beaver-overpopulation-tierra-del-fuego/.

34. See, e.g., Pablo Jusim, Andrea P. Goijman, Julio Escobar, María Luisa Carranza, and Adrián Schiavini, "First Test for Eradication of Beavers (*Castor canadensis*) in Tierra del Fuego, Argentina," *Biological Invasions* 22 (2020): 3609–3619.

35. Prabhat Kumar Rai, "Environmental Degradation by Invasive Alien Plants in the Anthropocene," *Anthropological* Science 1, no. 1 (2022): 5–28.

36. Invasive Species Advisory Committee, *Invasive Species Definition Clarification and Guidance*, April 27, 2006, https://www.doi.gov/sites/doi.gov/files/uploads/isac_definitions_white_paper_rev.pdf.

37. Calculated from numbers listed in Christopher L. Dutton, Amanda L. Subalusky, Stephen K. Hamilton, Emma J. Rosi, and David M. Post, "Organic Matter Loading by Hippopotami Causes Subsidy Overload Resulting in Downstream Hypoxia and Fish Kills," *Nature Communications* 9 (2018): 1951.

38. Amanda L. Subalusky, Elizabeth P. Anderson, Germán Jiménez, David M. Post, David Echeverri Lopez, Sebastián García-R, et al., "Potential Ecological and Socio-Economic Effects of a Novel Megaherbivore Introduction: The Hippopotamus in Colombia," *Oryx* 55, no. 1 (2021): 105–113; and Jonathan B. Shurin, Nelson Aranguren-Riaño, Daniel Duque Negro, David Echeverri Lopez, Natalie T. Jones, Oscar Laverde-R, et al., "Ecosystem Effects of the World's Largest Invasive Animal," *Ecology* 101, no. 5 (2020): e02991.

39. See Christopher B. Anderson, Guillermo Martínez Pastur, María Vanessa Lencinas, Petra K. Wallem, Michelle C. Moorman, and Amy D. Rosemond, "Do Introduced North American Beavers *Castor canadensis* Engineer Differently in Southern South America? An Overview with Implications for Restoration," *Mammal Review* 39, no. 1 (2009): 33–52; Gilliland, "Argentina Brought Beavers"; Gerardo E. Soto, Pablo M. Vergara, Marlene E. Lizama, Cristian Celis, Ricardo Rozzi, Quiterie Duron, et al., "Do Beavers Improve the Habitat Quality for Magellanic Woodpeckers?," *Bosque (Valdivia)* 33, no. 3 (2012): 271–274; and Ivan Arismendi, Brooke E. Penaluna, and Carlos G. Jara, "Introduced Beaver Improve Growth

of Non-Native Brown Trout in Tierra del Fuego, South America," *Ecology and Evolution* 10, no. 17 (2020): 9454–9465.

40. Ed Yong, "Hippos Poop So Much That Sometimes All the Fish Die," *Atlantic*, May 16, 2018, https://www.theatlantic.com/science/archive/2018/05/hippos-poop-so-much-that-sometimes-all-the-fish-die/560486/.

41. Dutton et al., "Organic Matter Loading by Hippopotami."

42. Matthew K. Chew, "The Monstering of the Tamarisk: How Scientists Made a Plant into a Problem," *Journal of the History of Biology* 42, no. 2 (2009): 231–266; and Candace Hughes, "How We Created a Monster in the Southwest," *Smithsonian Magazine*, October 9, 2015, https://www.smithsonianmag.com/science-nature/how-we-created-monster-american-southwest-180956878/.

43. Chew, "The Monstering of the Tamarisk," 231–266.

44. The same appears to be true for kudzu in the American South. Not only is the vine not quite as widespread as feared, but it also seems to thrive mainly in human-modified (and highly visible) road- and rail-side lots. See Bill Finch, "The True Story of Kudzu, the Vine That Never Ate the South," *Smithsonian Magazine*, September 2015, https://www.smithsonianmag.com/science-nature/true-story-kudzu-vine-ate-south-180956325/. Some readers might notice a parallel to Mary Shelley's *Frankenstein*, which, as Langdon Winner observed, is about "the plight of things that have been created but not in the context of sufficient care." The being that young Dr. Frankenstein brings to life is not born a monster but rather made into one. See Langdon Winner, *Autonomous Technology: Technics-out-of-Control as a Theme in Political Thought* (Cambridge, MA: MIT Press, 1978), 313.

45. Fred Pearce, *The New Wild: Why Invasive Species Will Be Nature's Salvation* (Boston: Beacon Press, 2015), 30–35.

46. Paul A. M. van Zwieten, Jeppe Kolding, Michael J. Plank, Robert E. Hecky, Thomas B. Bridgeman, Sally MacIntyre, et al., "The Nile Perch Invasion in Lake Victoria: Cause of Consequence of the Haplochromine Decline?," *Canadian Journal of Fisheries and Aquatic Sciences* 73, no. 4 (2016): 622–643; and Brian E. Marshall, "Guilty as Charged: Nile Perch Was the Cause of the Haplochromine Decline in Lake Victoria," *Canadian Journal of Fisheries and Aquatic Sciences* 75, no. 9 (2018): 1542–1559.

47. Asher Elbein, "Death Valley's Park Service Wants Them Gone. But Are the Wild Donkeys Really the Enemy?," *Undark*, October 14, 2019, https://undark.org/2019/10/14/death-valley-burros-fate/; and Erick J. Lundgren, "Equids Engineer Desert Water Availability," *Science* 371, no. 6541 (2021): 491–495.

48. Subalusky et al., "Potential Ecological and Socio-Economic Effects"; and Shurin et al., "Ecosystem Effects of the World's Largest Invasive Animal."

49. Tyler J. Kelley, "Navigating a Sea of Superlatives in Pursuit of the Asian Carp," *Undark*, July 23, 2018, https://undark.org/2018/06/25/asian-carp-chicago-illinois-lake-michigan/; Adam Hinterthuer, "The Explosive Spread of Asian Carp: Can the Great Lakes Be Protected?

Does It Matter?," *BioScience* 62, no. 3 (2012): 220–224; David Schaper, "Chicago Canal Flooded with Toxin to Kill Asian Carp," *NPR*, December 4, 2009, https://www.npr.org/2009/12/04/121104335/chicago-canal-flooded-with-toxin-to-kill-asian-carp; and Elizabeth Kolbert, *Under a White Sky: The Nature of the Future* (New York: Crown, 2021), 1–60.

50. Kolbert, *Under a White Sky*, 13-26; Hintertheur, "The Explosive Spread."

51. Some researchers dispute whether Japanese knotweed lives up to its reputation. Karen Bacon, "Japanese Knotweed Is No More of a Threat to Buildings Than Other Plants—New Study," *The Conversation*, July 9, 2018, https://theconversation.com/japanese-knotweed-is-no-more-of-a-threat-to-buildings-than-other-plants-new-study-99580.

52. Kupaa K. Luat-Hū'eu, Mehana Blaich Vaughan, and Melissa R. Price, "Understanding Local Pig Hunter Values and Practices as a Means Toward Co-Management of Feral Pigs (*Sus scrofa*; pua'a) in the Hawaiian Islands," *Ecology and Society* 28, no. 2 (2023): 32; Alexander Kelsch, Yasuo Takahashi, Rajarshi Dasgupta, André Derek Mader, Brian Alan Johnson, and Pankaj Kumar, "Invasive Alien Species and Local Communities in Social-Ecological Landscapes and Seascapes: A Systemic Review and Analysis," *Environmental Science and Policy* 112 (2020): 275–281.

53. CBS News, "Tamarisk Eradication Halted to Protect Endangered Bird," June 22, 2010, https://www.cbsnews.com/news/tamarisk-eradication-halted-to-protect-endangered-bird/. A similar dilemma arose in the San Francisco area, when it was recognized that the endangered clapper rail had come to rely on non-native cordgrass for habitat. Adam Lampert, Alan Hastings, Edwin D. Grosholz, Sunny L. Jardine, and James N. Sanchirico, "Optimal Approaches for Balancing Invasive Species Eradication and Endangered Species Management," *Science* 344, no. 6187 (2014): 1028–1031.

54. Martin A. Schlaepfer, Dov F. Sax, and Julian D. Olden, "The Potential Conservation Value of Non-Native Species," *Conservation Biology* 25, no. 3 (2011): 428–437; Jennifer Carpenter, "Alien Rats Take On Prey's Role," *BBC News*, November 16, 2011, https://www.bbc.com/news/science-environment-15738837; and Juan Carlos Castilla, Nelson A. Lagos, and Mauricio Cerda, "Marine Ecosystem Engineering by the Alien Ascidian *Pyrua praeputialis* on a Mid-Intertidal Rocky Shore," *Marine Ecology Progress Series* 258 (2004): 119–130.

55. See Justin P. Suraci, Kaitlyn M. Gaynor, Maximilian L. Allen, Peter Alexander, Justin S. Brashares, Sara Cendejas-Zarelli, et al., "Disturbance Type and Species Life History Predict Mammal Responses to Humans," *Global Change Biology* 27, no. 6 (2021): 3718–3731.

56. Zach St. George, "The Tree That's Rare, Endangered and Common," *Bay Nature*, June 21, 2020, https://baynature.org/article/the-tree-thats-rare-endangered-and-common/.

57. Shoshi Parks, "How the U.S. Almost Became a Nation of Hippo Ranchers," *Smithsonian Magazine*, May 30, 2023, https://www.smithsonianmag.com/history/how-the-us-almost-became-a-nation-of-hippo-ranchers-180982244/; and Alejandra Osejo Varona, "Hippos in Colombia: What to Do with Them?," *Rice University News*, November 18, 2022, https://graduate.rice.edu/news/current-news/hippos-colombia-what-do-them.

58. Christopher E. Doughty, Joe Roman, Søren Faurby, and Jens-Christian Svenning, "Global Nutrient Transport in a World of Giants," *Proceedings of the National Academic of Science* 113, no. 4 (2015): 868–873; Amanda L. Subalusky, Christopher L. Dutton, Emma J. Rosi-Marshall, and David M. Post, "The Hippopotamus Conveyor Belt: Vectors of Carbon and Nutrients from Terrestrial Grasslands to Aquatic Systems in Sub-Saharan Africa," *Freshwater Biology* 60, no. 3 (2015): 512–525; and Jacob Dembitzer, "The Case for Hippos in Colombia," *Israel Journal of Ecology & Evolution* 63, no. 3–4 (2017): 5–8.

59. Daniel Simberloff, *Invasive Species: What Everyone Needs to Know* (New York: Oxford University Press, 2013).

60. Montserrat Vilà, Corina Basnou, Petr Pyšek, Melanie Josefsson, Piero Genovesi, Stephan Gollasch, et al., "How Well Do We Understand the Impacts of Alien Species on Ecosystem Services?," *Frontiers in Ecology and the Environment* 8, no. 3 (2010): 135–144.

61. Jonathan M. Jeschke and Petr Pyšek, "Tens Rule," in *Invasion Biology: Hypothesis and Evidence*, ed. Jonathan M. Jeschke and Tina Heger (Boston, MA: CABI, 2018), 124–132.

62. Manuel-Angel Dueñas, David J. Hemming, Amy Roberts, and Hilda Diaz-Soltero, "The Threat of Invasive Species to IUCN-Listed Critically Endangered Species: A Systematic Review," *Global Ecology and Conservation* 26 (2021): e01476.

63. Brianna Randall, "America's Billion-Dollar Tree Problem Is Spreading," *Wired*, November 23, 2022, https://www.wired.com/story/billion-dollar-encroaching-tree-problem/.

64. Truman P. Young, Duncan M. Kimuyu, Wilfred O. Odadi, Harry B. M. Wells, and Amelia A. Wolf, "Naïve Plant Communities and Individuals May Initially Suffer in the Face of Reintroduced Megafauna: An Experimental Exploration of Rewilding from an African Savanna Rangeland," *PLoS One* 16, no. 4 (2021): e0248855, https://doi.org/10.1371/journal.pone.0248855.

65. One review finds a surprisingly high level of biodiversity and provision of ecosystem services among novel ecosystems. Cody R. Evers, Chloe B. Wardropper, Ben Branoff, Elise F. Granek, Shana L. Hirsch, Timothy E. Link, et al., "The Ecosystem Services and Biodiversity of Novel Ecosystems," *Global Ecology and Conservation* 13 (2018): e00362.

66. Emma Marris, "Ragamuffin Earth," *Nature* 460, no. 23 (2009): 450–453.

67. Marris, "Ragamuffin Earth," 450–453; Ariel E. Lugo, "The Outcome of Alien Tree Invasions in Puerto Rico," *Frontiers in Ecology and the Environment* 2, no. 5 (2004): 265–273; Joseph Mascaro, R. Flint Hughes, and Stefan A. Schnitzer, "Novel Forests Maintain Ecosystem Processes After the Decline of Native Tree Species," *Ecological Monographs* 82, no. 2 (2012): 221–228; Marris, *Rambunctious Garden*, 111–122; and Scott Dance and Kate Selig, "Why Hawaii's Wildfires Are So Devastating—and 'Predictable,'" *Washington Post*, August 10, 2023, https://www.washingtonpost.com/weather/2023/08/10/maui-wildfires-why-weather-climate/.

68. Mary Douglas and Aaron Wildavsky, *An Essay on the Selection of Technological and Environmental Dangers* (Berkeley: University of California Press, 1982).

69. Linda Step and Inge Sievers, "Cultural Theory and Individual Perceptions of Environmental Risks," *Environment and Behavior* 32, no. 2 (2000): 250–269.

70. Author creation. Marble metaphor adapted from Step and Sievers, "Cultural Theory and Individual Perceptions," 254.

71. Ben L. Phillips and Richard Shine, "The Morphology, and Hence Impact, of an Invasive Species (the Cane Toad, *Bufo marinus*): Changes with Time Since Colonization," *Animal Conservation* 8, no. 4 (2005): 407–413; Shine and Wiens, "The Ecological Impact of Invasive Cane Toads"; Ben L. Phillips and Richard Shine, "An Invasive Species Induced Rapid Adaptive Change in a Native Predator: Cane Toads and Black Snakes in Australia," *Proceedings of the Royal Society B: Biological Sciences* 273, no. 1593: 1545–1550; and Rick Shine, *Cane Toad Wars* (Berkeley: University of California Press, 2018), 128–129, 233.

72. Simberloff, *Invasive Species*, 238; and Jeffrey T. Foster and Scott K. Robinson, "Introduced Birds and the Fate of Hawaiian Rainforests," *Conservation Biology* 21, no. 5 (2007): 1248–1257.

73. Daniel Simberloff and Betsy Von Holle, "Positive Interactions of Nonindigenous Species: Invasional Meltdown?," *Biological Invasions* 1 (1999): 21–32.

74. Ben J. Novak, Ryan Phelan, and Michele Weber, "U.S. Conservation Translocations: Over a Century of Intended Consequences," *Conservation Science and Practice* 3, no. 4 (2021): e394.

75. Stephenie Livingston, "Meet the World's Deadliest Duo," *University of Florida News*, January 3, 2017, https://news.ufl.edu/articles/2017/01/meet-the-worlds-deadliest-duo.html.

76. Scott P. Carroll, "Conciliation Biology: The Eco-Evolutionary Management of Permanently Invaded Biotic Systems," *Evolutionary Applications* 4, no. 2 (2011): 184–199.

77. Christopher J. Jolly, "Out of the Frying Pan: Reintroduction of Toad-Smart Northern Quolls to Southern Kakadu National Park," *Austral Ecology* 43, no. 2 (2018): 139–149.

78. Dembitzer, "The Case for Hippos in Colombia."

79. Nicholas J. Reo and Laura A. Ogden, "Anishnaabe: Aki: An Indigenous Perspective on the Global Threat of Invasive Species," *Sustainability Science* 13 (2018): 143–1452.

80. Thomas, *Inheritors of the Earth*, 124–126.

81. Quoted in Elbein, "Death Valley's Park Service."

82. Anthony Ricciardi, "Ballast Water Management Is Reducing the Flow of Invasive Species into the Great Lakes," *The Conversation*, October 12, 2022, https://theconversation.com/ballast-water-management-is-reducing-the-flow-of-invasive-species-into-the-great-lakes-190880.

83. The situation, even for chytrid and frogs, isn't black and white. Some frog species are developing a resistance to the fungus, which is enabling selective breeding and reintroduction programs. Moreover, heat-loving frogs can be assisted in their recovery from the fungus by placing large bricks with frog-sized holes in their habitat. By sheltering in these structures, the animals are far more likely to survive. Martin J. Kernan, "Frogs Kick Back Against

Lethal Fungus," *Knowable Magazine*, November 6, 2024, https://knowablemagazine.org/content/article/living-world/2024/new-strategies-offer-hope-for-frogs-dying-from-fungus.

84. Christopher Joyce, "How the Smokey Bear Effect Led to Raging Wildfires," *NPR*, August 23, 2012, https://www.npr.org/2012/08/23/159373691/how-the-smokey-bear-effect-led-to-raging-wildfires.

85. Escobar, *Pablo Escobar.*

86. Diana Durán, "Colombia Tries a New Solution for Pablo Escobar's Cocaine Hippos," *Washington Post*, March 3, 2023, https://www.washingtonpost.com/world/2023/03/03/colombia-escobar-hippos/.

87. Marko Álvarez and Astrid Suárez, "'No Los Maten': Así Defienden a Hipopótamos de Pablo Escobar," *AP News*, February 19, 2022, https://apnews.com/article/87eec95fe7fd4836458b1675dd38f83f/. Author translation of quote.

88. Shurin et al., "Ecosystem Effects of the World's Largest Invasive Animal," 2.

89. Matthew Chew describes how the claim that invasive species are the "second greatest threat" to biodiversity was founded on shaky scientific grounds, a product of E. O. Wilson's selective curation of conservationists' apocalyptic concerns. See Matthew K. Chew, "Ecologists, Environments, Experts, and the Invasion of the 'Second Greatest Threat,'" *International Review of Environmental History* 1 (2015): 17–40.

90. Thomas, *Inheritors of the Earth*, 114.

91. David M. Frank, "Disagreement or Denialism? 'Invasive Species Denialism' and Ethical Disagreement in Science," *Synthese* 198, Suppl. 25 (2021): S6104.

CHAPTER 4

1. Joel K. Bourne Jr., "In the Arctic's 'Cold Rush,' There Are No Easy Profits," *National Geographic*, March 1, 2016, https://www.nationalgeographic.com/magazine/article/new-arctic-thawing-rapidly-circle-work-oil/.

2. David Kindy, "Permafrost Thaw in Siberia Creates a Ticking 'Methane Bomb' of Greenhouse Gases, Scientists Warn," *Smithsonian Magazine*, August 5, 2021, https://www.smithsonianmag.com/smart-news/ticking-timebomb-siberia-thawing-permafrost-releases-more-methane-180978381/.

3. Sergey A. Zimov, "Pleistocene Park: Return of the Mammoth's Ecosystem," *Science* 308, no 5723 (2005): 796–798; and Paul Jepson and Cain Blythe, *Rewilding: The Radical New Science of Ecological Recovery* (London: Icon, 2020), 51, 82.

4. Zimov, "Pleistocene Park."

5. Paul S. Martin, *Twilight of the Mammoths* (Berkeley: University of California Press, 2005).

6. Daniel H. Mann, Pamela Groves, Benjamin V. Gaglioti, and Beth A. Shapiro, "Climate-Driven Ecological Stability as the Globally Shared Cause of Late Quaternary Megafaunal Extinctions," *Biological Reviews* 94, no. 1 (2018): 328–352.

7. Hannah Ritchie, "Did Humans Cause the Quaternary Megafauna Extinction?," *Our World in Data*, November 30, 2022, https://ourworldindata.org/quaternary-megafauna-extinction#:~:text=Between%2050%20and%2040%2C000%20years,North%20and%20South%20America%20occurred.

8. Jepson and Blythe, *Rewilding*, 30–31.

9. Anthony John Stuart, "Late Quaternary Megafaunal Extinctions on the Continents: A Short Review," *Geological Journal* 50, no. 3 (2014): 338–363.

10. Christophere Sandom, Søren Faurby, Brody Sandel, and Jens-Christian Svenning, "Global Late Quaternary Megafauna Extinctions Linked to Humans, Not Climate Change," *Proceedings of the Royal Society B* 281, no. 1787 (2014): 20133254.

11. Lisa Nagaoka, Torben Rick, and Steve Wolverton, "The Overkill Model and Its Impact on Environmental Research," *Ecology and Evolution* 8, no. 19 (2018): 9683–9696.

12. Ross Anderson, "Welcome to Pleistocene Park," *Atlantic*, April 2017, https://www.theatlantic.com/magazine/archive/2017/04/pleistocene-park/517779/; and *Pleistocene Park*, https://pleistocenepark.ru//.

13. Zimov, "Pleistocene Park"; and Anderson, "Welcome to Pleistocene Park."

14. Sergey Zimov, "The Woolly Mammoth Revival," *Revive & Restore*, November 25, 2014, https://reviverestore.org/projects/woolly-mammoth/sergey-zimovs-manifesto//.

15. David M. Hopkins, John V. Matthews, Charles E. Schweger, David M. Hopkins, John V. Matthews, and Charles E. Schweger, eds., *Paleoecology of Beringia* (New York: Academic Press, 1982).

16. Gregory P. Asner and Shaun R. Levick, "Landscape-Scale Effects of Herbivores on Treefall in African Savannas," *Ecology Letters* 15 (2012): 1211–1217.

17. Sergey A. Zimov, N. S. Zimov, A. N. Tikhonov, and F. S. Chapin III, "Mammoth Steppe: A High-Productivity Phenomenon," *Quaternary Science Reviews* 57, no. 4 (2012): 26–45.

18. Christian Beer, Nikita Zimov, Johan Olofsson, Philipp Porada, and Sergey Zimov, "Protection of Permafrost Soils from Thawing by Increasing Herbivore Density," *Nature Scientific Reports* 10, no. 4170 (2020), https://doi.org/10.1038/s41598-020-60938-y.

19. William Cronon, "The Trouble with Wilderness: Or, Getting Back to the Wrong Nature," *Environmental History* 1, no. 1 (1996): 7–28.

20. Dave Foreman, "Around the Campfire," *Wild Earth* 2, no. 3 (1992): inside front cover.

21. Michael E. Soulé and Reed Noss, "Rewilding and Biodiversity: Complementary Goals for Continental Conservation," *Wild Earth* 8, no. 3 (1998), 18–28; and Dolly Jørgensen, "Rethinking Rewilding," *Geofourm* 65 (2015): 482–488.

22. Josh Donlan, "Re-wilding North America," *Nature* 436 (2005): 913–914.

23. Dave Foreman, *Confessions of an Eco-Warrior* (New York: Crown, 1991), 69.

24. Christopher N. Johnson, Lynda D. Prior, Sally Archibald, Helen M. Poulos, Andrew M. Barton, and Grant J. Williamson, "Can Trophic Rewilding Reduce the Impact of Fire in

a More Flammable World?," *Philosophical Transactions of the Royal Society B* 373, no. 1761 (2018): 20170443.

25. See Doughty et al., "Global Nutrient Transport"; and James Ashworth, "Artificial Whale Poo Could Help Restore Ocean Biodiversity," *The Natural History Museum London*, March 9, 2022, https://www.nhm.ac.uk/discover/news/2022/march/artificial-whale-poo-could-help-restore-ocean-biodiversity.html/.

26. Nathalie Pettorelli, Sarah M. Durant, and Johan T. Du Toit, "Rewilding: A Captivating, Controversial, Twenty-First-Century Concept to Address Ecological Degradation in a Changing World," in *Rewilding*, ed. Nathalie Pettorelli, Sarah M. Durant, and Johan T. Du Toit (New York: Cambridge University Press, 2019), 1–11.

27. Roderick F. Nash, *Wilderness and the American Mind* (New Haven, CT: Yale University Press, 2014), xx.

28. Pettorelli et al., "Rewilding," 7. Andrea Perino et al. define this self-sustaining ecosystem in terms of trophic complexity, the dominance of random environmentally driven disturbances, and high levels of connectivity between habitats to enable dispersal. Andrea Perino, Henrique M. Pereira, Laetitia M. Navarro, Néstor Fernández, James M. Bullock, Silvia Ceauşu, et al., "Rewilding Complex Ecosystems," *Science* 364, no. 6438 (2019): eaav5570.

29. Kaya Klop-Toker, Simon Clulow, Craig Shuttleworth, and Matt W. Hayward, "Are Novel Ecosystems the Only Novelty of Rewilding?," *Restoration Ecology* 28, no. 6 (2020): 1318–1320.

30. David Nogués-Bravo, Daniel Simberloff, Carsten Rahbek, and Nathan James Sanders, "Rewilding Is the New Pandora's Box in Conservation," *Current Biology* 26, no. 3 (2016): R83–R101; and Dustin R. Rubenstein and Daniel I. Rubenstein, "From Pleistocene to Trophic Rewilding: A Wolf in Sheep's Clothing," *Proceedings of the National Academy of Sciences* 113, no. 1 (2016): E1.

31. Michael S. Carolan, "Conserving Nature, But to What End?," *Organization & Environment* 19, no. 2 (2006): 153–170.

32. Jens-Christian Svenning, Pil B. M. Pedersen, C. Josh Donlan, Rasmus Ejrnæs, Søren Faurby, Mauro Galetti, et al., "Reply to Rubenstein and Rubenstein: Time to Move on from Ideological Debates on Rewilding," *Proceedings of the National Academy of Sciences* 113, no. 1 (2016): E2–E3.

33. BBC News, "Tortoise Eating a Bird Caught on Camera—BBC News," YouTube, 1:54, August 29, 2012, https://www.youtube.com/watch?v=C0tjq0u2rDU/.

34. ScienceDaily, "Tortoise Evolution: How Did They Become So Big?," October 23, 2018, https://www.sciencedaily.com/releases/2018/10/181023110534.htm.

35. Christine J. Griffiths, Dennis M. Hansen, Carl G. Jones, Nicolas Zuël, and Stephen Harris, "Resurrecting Extinct Interactions with Extant Substitutes," *Current Biology* 21, no. 9 (2011): 762–765.

36. Griffiths et al., "Resurrecting Extinct Interactions," 762–765; and Rosemary J. Moorhouse-Gann, Ian P. Vaughan, Nik C. Cole, Martine Goder, Vikash Tatayah, Carl G. Jones, et al.,

"Impacts of Herbivory by Ecological Replacements on an Island Ecosystem," *Journal of Applied Ecology* 59, no. 9 (2021): 2245–2261.

37. Moorhouse-Gann et al., "Impacts of Herbivory," 2245–2261; and Annika Waibel, Christine J. Griffiths, Nicolas Zuël, Bernhard Schmid, and Matthias Albrecht, "Does a Giant Tortoise Taxon Substitute Enhance Seed Germination of Exotic Fleshy-Fruited Plants?," *Journal of Plant Ecology* 6, no. 1 (2013): 57–63.

38. Erica Gies, "Unique Elk in California May Be Killed Under Controversial Plan," *National Geographic*, September 30, 2020, https://www.nationalgeographic.com/animals/article/tule-elk-culled-under-point-reyes-proposal/.

39. Brent E. Johnson and J. Hall Cushman, "Influence of a Large Herbivore Reintroduction on Plant Invasions and Community Composition in a California Grassland," *Conservation Biology* 21, no. 2 (2007): 515–526; and Joshua Rapp Learn, "It Might Be Impossible to Turn Back the Clock on Altered Ecosystems," *Smithsonian Magazine*, March 8, 2016, https://www.smithsonianmag.com/science-nature/it-might-be-impossible-turn-back-clock-altered-ecosystems-180958316//.

40. Lacey F. Hughey, Kevin T. Shoemaker, Kelley M. Stewart, Douglas J. McCauley, and J. Hall Cushman, "Effects of Human-Altered Landscapes on a Reintroduced Ungulate," *Biological Conservation* 257 (2021): 109086.

41. National Park Service, "Ranching History at Point Reyes," March 24, 2023, https://www.nps.gov/pore/learn/historyculture/stories_ranching.htm/.

42. Gies, "Unique Elk in California."

43. James Steinbauer, "On the California Coast, It's Cattle Versus Elk," *Sierra*, October 7, 2019, https://www.sierraclub.org/sierra/california-coast-it-s-cattle-versus-elk.

44. Resource Renewal Institute, "Tule Elk: The Killing of a Native Species," Vimeo, 8:21, September 17, 2020, https://vimeo.com/459015215.

45. Learn, "It Might Be Impossible."

46. Learn, "It Might Be Impossible."

47. Sustainable Human, "How Wolves Change Rivers," YouTube, 4:33, February 13, 2014, https://www.youtube.com/watch?v=W88Sact1kws/.

48. Jens-Christian Svenning, Michael Munk, and Andreas Schweiger, "Trophic Rewilding," in *Rewilding*, ed. Nathalie Pettorelli, Sarah M. Durant, and Johan T. Du Toit (New York: Cambridge University Press, 2019), 73–98.

49. Frederic H. Wagner, "Scientists Say Yellowstone Park Is Being Destroyed," *High Country News*, May 30, 1994, https://www.hcn.org/issues/12/370.

50. Emma Marris, "A Good Story: Media Bias in Trophic Cascade Research in Yellowstone National Park," in *Effective Conservation Science: Data not Dogma*, ed. Peter Kareiva, Michelle Marvier, and Brian Silliman (New York: Oxford University Press, 2017), 80–84.

51. Rolf O. Peterson, John A. Vucetich, Joseph M. Bump, and Douglas W. Smith, "Trophic Cascades in a Multicausal World," *Annual Review of Ecology, Evolution, and Systematics* 45, no. 1 (2014): 325–345.

52. Matt W. Hayward, Sarah Edwards, Bronwyn A. Fancourt, John D. C. Linnell, and Erlend B. Nilsen, "Top-Down Control of Ecosystems and the Case for Rewilding? Does It All Add Up?," in *Rewilding*, ed. Nathalie Pettorelli, Sarah M. Durant, and Johan T. Du Toit (New York: Cambridge University Press, 2019), 325–354.

53. Daniel B. Botkin, *Moon in the Nautilus Shell: Discordant Harmonies Reconsidered* (New York: Oxford University Press, 2012).

54. Paul Jepson, "Recoverable Earth: A Twenty-First Century Environmental Narrative," *Ambio* 48 (2019): 123–130.

55. Alex Spencer, "George Monbiot: Farming Is the Number One Threat to the Planet—And We Must Switch to Food Produced by Bacteria," *Cambridge Independent*, November 6, 2022, https://www.cambridgeindependent.co.uk/whats-on/george-monbiot-farming-is-the-number-one-threat-to-the-plan-9280753/; and George Monbiot, "Lab-Grown Food Will Soon Destroy Farming—And Save the Planet," *Guardian*, January 8, 2020, https://www.theguardian.com/commentisfree/2020/jan/08/lab-grown-food-destroy-farming-save-planet.

56. Jeff Vance Martin, "In the Shadow of the Wolf: Wildlife Conflict and Land Use Politics in the New West" (PhD diss., University of California, Berkeley, 2020).

57. Steph Yin, "The Netherlands' Grand Rewilding Experiment, Gone Haywire," *WHYY*, August 23, 2019, https://whyy.org/segments/the-netherlands-grand-rewilding-experiment-gone-haywire/.

58. Yin, "The Netherlands' Grand Rewilding Experiment"; and Elizabeth Kolbert, "Recall of the Wild," *New Yorker*, December 16, 2012, https://www.newyorker.com/magazine/2012/12/24/recall-of-the-wild.

59. E. M. Kintzel and I. van Stokkum, "Konik-Pferde in den Oostvaardersplassen," *Wikimedia*, July 25, 2008, https://commons.wikimedia.org/wiki/File:Oostvaardersplassen_koniks.jpg, used under Attribution-ShareAlike 3.0 Unported, https://creativecommons.org/licenses/by-sa/3.0/deed.en.

60. Lorraine Boissoneault, "When the Nazis Tried to Bring Animals Back from Extinction," *Smithsonian Magazine*, March 31, 2017, https://www.smithsonianmag.com/history/when-nazis-tried-bring-animals-back-extinction-180962739/.

61. Frans W. M. Vera, "Large-Scale Nature Development—The Oostvaardersplassen," *British Wildlife*, June (2009): 28–36.

62. Jepson and Blythe, *Rewilding*, 45–48; and Frans W. M. Vera, *Grazing Ecology and Forest History* (New York: CABI, 2000).

63. Jepson and Blythe, *Rewilding*, 49; and Jamie Lorimer, *Wildlife in the Anthropocene* (Minneapolis: University of Minnesota Press, 2015), 105.

64. Lorimer, *Wildlife in the Anthropocene*, 107; and Yin, "The Netherlands' Grand Rewilding Experiment."

65. Patrick Barkham, "Dutch Rewilding Experiment Sparks Backlash as Thousands of Animals Starve," *Guardian*, April 27, 2018, https://www.theguardian.com/environment/2018/apr/27/dutch-rewilding-experiment-backfires-as-thousands-of-animals-starve; and Lorimer, *Wildlife in the Anthropocene*, 110.

66. Vera, "Large-Scale Nature Development," 35.

67. Kolbert, "Recall of the Wild."

68. Helen Kopina, Simon Leadbeater, and Paul Cryer, "The Golden Rules of Rewilding—Examining the Case of Oostvaardersplassen," *ECOS* 40, no. 6 (2019), https://www.ecos.org.uk/ecos-406-the-golden-rules-of-rewilding-examining-the-case-of-oostvaardersplassen/.

69. Kopina et al., "The Golden Rules."

70. Martha C. Nussbaum, "A Peopled Wilderness," *New York Review of Books*, December 8, 2022, https://www.nybooks.com/articles/2022/12/08/a-peopled-wilderness-martha-c-nussbaum/.

71. Yin, "The Netherlands' Grand Rewilding Experiment."

72. *Chernobyl Reclaimed: An Animal Takeover*, directed by Peter Hayden (Off the Fence, 2007), 50:30. https://therokuchannel.roku.com/watch/0966acb8f38229a6372cc455f3234da2.

73. "Abandoned Buildings in Chernobyl," by Gáspár Ferenc, *pexels*, April 28, 2018, https://www.pexels.com/photo/abandoned-buildings-in-chernobyl-13714462/.

74. Adam Higginbotham, "Is Chernobyl a Wild Kingdom or a Radioactive Den of Decay?," *Wired*, April 14, 2011, https://www.wired.com/2011/04/ff-chernobyl/; and Timothy A. Mousseau, "The Biology of Chernobyl," *Annual Review of Ecology, Evolution, and Systematics* 52, no. 1 (2021): 87–109.

75. Valery C. Dombrovski, Dzmitry V. Zhurauliou, and Adham Ashton-Butt, "Long-Term Effects of Rewilding on Species Composition," *Restoration Ecology* 30, no. 8 (2022): e13633; Sarah C. Webster, Michael E. Byrne, Stacey L. Lance, Cara N. Love, Thomas G. Hinton, Dmitry Shamovich, et al., "Where the Wild Things Are: Influence of Radiation on the Distribution of Four Mammalian Species Within the Chernobyl Exclusion Zone," *Frontiers in Ecology and the Environment* 14, no. 4 (2016): 186–190; Germán Orizaola, "Chernobyl Has Become a Refuge for Wildlife 33 Years After the Nuclear Accident," *The Conversation*, May 8, 2019, https://theconversation.com/chernobyl-has-become-a-refuge-for-wildlife-33-years-after-the-nuclear-accident-116303; and Tania Rabesandratana, "Humans Are Worse Than Radiation for Chernobyl Animals, Study Finds," *Science*, October 5, 2015, https://www.science.org/content/article/humans-are-worse-radiation-chernobyl-animals-study-finds/.

76. Kelly Servick, "Is a Little Radiation Good for You?," *Science*, October 4, 2018, https://www.science.org/content/article/little-radiation-good-you-controversial-theory-pops-senate-hearing-epa-transparency.

77. Higginbotham, "Is Chernobyl a Wild Kingdom."

78. Alistair Walsh, "Passive Rewilding: 'If You Love It, Set It Free,'" *Deutsche Welle*, November 11, 2021, https://www.dw.com/en/the-case-for-passive-rewilding-if-you-love-it-let-it-free/a-59450781; and Perino et al., "Rewilding Complex Ecosystems."

79. Perino et al., "Rewilding Complex Ecosystems"; Euan N. Furness, Russell J. Garwood, Philip D. Mannion and Mark D. Sutton, "Evolutionary Simulations Clarify and Reconcile Biodiversity-Disturbance Models," *Proceedings of the Royal Society B* 288, no. 1949 (2021): 20210240.

80. Laetitia M. Navarro and Henrique Pereira, "Rewilding Abandoned Landscapes in Europe," *Ecosystems* 15 (2012): 900–912.

81. Cristina Quintas-Soriano, "Impacts of Land Use Changes on Ecosystem Services and Implications for Human Well-Being in Spanish Drylands," *Land Use Policy* 54 (2016): 534–548; Emma H. van der Zanden, Sonia M. Carvalho-Ribeiro, and Peter H. Verberg, "Abandonment Landscapes: User Attitudes, Alternative Futures and Land Management in Castro Laboreiro, Portugal," *Regional Environmental Change* 18, no. 5 (2018): 1509–1520; and Rubén Granado-Díaz, Anastasio J. Villanueva, and José A. Gomez-Limón, "Willingness to Accept for Rewilding Farmland in Environmentally Sensitive Areas," *Land Use Policy* 116 (2022): 105052.

82. Iago Otero, Joan Marull, Enric Tello, Giovanna L. Diana, Manel Pons, Francesc Coll, et al., "Land Abandonment, Landscape, and Biodiversity: Questioning the Restorative Character of the Forest Transition in the Mediterranean," *Ecology and Society* 20, no. 2 (2015), http://www.jstor.org/stable/26270180; Chieko Koshida and Naoki Katayama, "Meta-analysis of the Effects of Rice-Field Abandonment on Biodiversity in Japan," *Conservation Biology* 32, no. 6 (2018): 1392–1402; and Taku Kadoya and Izumi Washitani, "The Satoyama Index: A Biodiversity Indicator for Agricultural Landscapes," *Agriculture, Ecosystems & Environment* 140, no. 1–2 (2011): 20–26.

83. Marc Macias-Fauria, Paul Jepson, Nikita Zimov, and Yadvinder Malhi, "Pleistocene Arctic Megafaunal Ecological Engineering as a Natural Climate Solution?," *Philosophical Transactions of the Royal Society London B: Biological Sciences* 375, no. 1794 (2020): 20190122; and Freda Kreier, "Russia's War in Ukraine Is Disrupting Studies of Ancient Life," *Nature*, June 30, 2022, https://www.nature.com/articles/d41586-022-01790-0.

84. Wolfgang Fischer, Christoph K. Thomas, Nikita Zimov, and Mathias Göckede, "Grazing Enhances Carbon Cycling but Reduces Methane Emission During Peak Growing Season in the Siberian Pleistocene Park Tundra Site," *Biogeosciences* 19, no. 6 (2022): 1611–1633.

85. Cronon, "The Trouble with Wilderness," 24.

86. Marris, *Rambunctious Garden*, quoted in Anderson, "Welcome to Pleistocene Park." Jamie Lorimer develops these themes further in his book, *Wildlife in the Anthropocene*, at least in a way that some academic readers might find theoretically rich and insightful. From what I can tell, Lorimer and I are feeling our way toward similar perspectives on the Anthropocene. See Lorimer, *Wildlife in the Anthropocene.*

CHAPTER 5

1. Hannah Nordhaus, "Two Visions Collide amid Push to Restore Montana Plains," *National Geographic*, January 16, 2020, https://www.nationalgeographic.com/magazine/article/two-visions-collide-amid-push-to-restore-montana-plain-feature; and Shawn Regan, "Where the Buffalo Roam," *Breakthrough Journal* 10 (2019), https://thebreakthrough.org/journal/no-10-winter-2019/where-the-buffalo-roam.
2. Nordhaus, "Two Visions Collide."
3. Regan, "Where the Buffalo Roam."
4. Michael J. Dax, *Grizzly West: A Failed Attempt to Reintroduce Grizzly Bears in the Mountain West* (Lincoln: University of Nebraska Press, 2015), 158.
5. Dax, *Grizzly West*, 17–44.
6. Regan, "Where the Buffalo Roam"; and Martin, "In the Shadow of the Wolf," 45–46, 81.
7. John Newton, "Farm Contribution to Agricultural GDP at Record Low," *Farm Bureau*, June 12, 2018, https://www.fb.org/market-intel/farm-contribution-to-agricultural-gdp-at-record-low; and "Outdoor Recreation Satellite Account, U.S. and States, 2021," *Bureau of Economic Analysis*, November 9, 2022, https://www.bea.gov/news/2022/outdoor-recreation-satellite-account-us-and-states-2021.
8. Dax, *Grizzly West*, 1–16, 165–186; and Martin, "In the Shadow of the Wolf," 7.
9. Dax, *Grizzly West*, 76–77, 91–114.
10. Dax, *Grizzly West*, 122, 125–127.
11. Dax, *Grizzly West*, 195.
12. United States Department of Agriculture, *Death Loss in U.S. Cattle and Calves Due to Predator and Nonpredator Causes, 2015*, December 2017, https://www.aphis.usda.gov/animal_health/nahms/general/downloads/cattle_calves_deathloss_2015.pdf; and Martin, "In the Shadow of the Wolf," 80.
13. Dax, *Grizzly West*, 206, 213–214.
14. Dax, *Grizzly West*, 213, 222.
15. Sonja K. Pieck, "Conserving Novel Ecosystems and Layered Landscapes Along the Inter-German Border," *Landscape Research* 45, no. 3 (2020): 346–358.
16. Tay Wiles and Brooke Warren, "Federal-Lands Ranching: A Half-Century of Decline," *High Country News*, June 13, 2016; and Nordhaus, "Two Visions Collide"; "APR's Comprehensive Plan to Destroy Western Ranching," *Rangefire!*, March 7, 2021, https://rangefire.us/2021/03/07/aprs-comprehensive-plan-to-destroy-western-ranching/.
17. Dax, *Grizzly West*, 38–40.
18. Damien M. Schiff, "The Endangered Species Act at 40," *Environs* 37 (2013–2014): 105–132; *Dax*, Grizzly West, 50; and Michael S. Carolan, "The Politics in Environmental Science: The

Endangered Species Act and the Preble's Mouse Controversy," *Environmental Politics* 17, no. 3 (2008): 449–465.

19. Schiff, "The Endangered Species Act," 112, n. 46. There is also an executive committee, referred to as the "God Squad," that technically has the authority to exempt FWS from having to act, but it has very rarely used its powers.

20. Dax, *Grizzly West*, 52. In fact, the authors of a 2025 study argue that the snail darter isn't actually a distinct species but rather a population of the stargazing darter. One of them contends that earlier scientists "squinted their eyes a bit" in order to gain a powerful tool against the Tellico Dam. Ava Ghezelayagh, Jeffrey W. Simmons, Julia E. Wood, Tsunemi Yamashita, Matthew R. Thomas, Rebecca E. Blanton, et al., "Comparative Species Delimitation of a Biological Conservation Icon," *Current Biology* 35, no. 2 (2025): 398–406.e4; and Jason Nark, "The Tiny Fish's Mistaken Identity Halted a Dam's Construction," *New York Times*, January 3, 2025, https://www.nytimes.com/2025/01/03/science/snail-darter-fish-tellico-dam.html.

21. "Preble's Meadow Jumping Mouse" by Rob Schorr/Colorado Natural Heritage Program, *flickr*, April 3, 2020, https://www.flickr.com/photos/usfwsmtnprairie/49730649486.

22. "Preble's Meadow Jumping Mouse (*Zapus Hudsonius Preblei*)," *U.S. Fish & Wildlife Service*, accessed July 20, 2023, https://ecos.fws.gov/ecp/species/4090.

23. Carolan, "The Politics in Environmental Science"; and Robert R. Crifasi, "A Subspecies No More? A Mouse, Its Unstable Taxonomy, and Wester Riparian Resource Conflict," *Cultural Geographies* 14 (2007): 511–535.

24. Carolan, "The Politics in Environmental Science," 456–457.

25. Carolan, "The Politics in Environmental Science," 457.

26. A panel convened in 2006 by FWS declared the Preble's mouse to be a unique species, siding with genetic analysis by Tim King. One interviewed scientist commented, "Both [sides] collected data relevant to the questions they posed and both addressed them in ways that they thought were reasonable and came to opposite conclusions . . . How do you resolve that? You don't. You argue until the cows come home." Christie Aschwanden, "Is It or Isn't It (Just Another Mouse)?," *High Country News*, August 7, 2006, https://www.hcn.org/issues/issue-327/is-it-or-isnt-it-just-another-mouse/.

27. See Crifasi, "A Subspecies No More?"

28. Dax, *Grizzly West*, 123.

29. Adam Liptak, "The Dusky Gopher Frog Loses a Round in the Supreme Court," *New York Times*, November 27, 2018, https://www.nytimes.com/2018/11/27/us/politics/supreme-court-gopher-frog.html; and Jonathan Wood and Tate Watkins, "Critical Habitat's Unique 'Private Land Problem:' Lessons from the Dusky Gopher Frog," *The Center for Growth and Opportunity*, Working Paper 2020.018, https://www.thecgo.org/research/critical-habitats-unique-private-land-problem/.

30. Terry Anderson, "When the Endangered Species Act Threatens Wildlife," *Hoover Institution*, October 20, 2014, https://www.hoover.org/research/when-endangered-species-act-threatens-wildlife.

31. Amara Brook, Michaela Zint, and Raymond De Young, "Landowners' Responses to an Endangered Species Act Listing and Implications for Encouraging Conservation," *Conservation Biology* 17, no. 6 (2003): 1638–1649.

32. Nordhaus, "Two Visions Collide."

33. Dave Skinner, "Critical Mass," *Range Magazine*, Fall 2019, 16–31.

34. Nicholas Lemann, "No People Allowed," *New Yorker*, November 14, 1999, https://archives.newyorker.com/?i=1999-11-22#folio=096.

35. Dax, *Grizzly West*, 60–62; Lowell E. Baier, *Inside the Equal Access to Justice Act* (New York: Rowman and Littlefield, 2016), 201–269; and Rich, *Getting to Green*, 31–61.

36. Baier, *Inside the Equal Access*, 37–92.

37. Baier, *Inside the Equal Access*, 271–303.

38. Baier, *Inside the Equal Access*, 271–303.

39. Michael Doyle and Pamela King, "How the Courthouse Came to Be a Major Habitat for the ESA," *Greewire*, June 20, 2023, https://www.eenews.net/articles/how-the-courthouse-came-to-be-a-major-habitat-for-the-esa/; and Dino Grandoni, "Want to Save the Whales? Reconsider the Lobster, Some Say," *Washington Post*, December 3, 2022, https://www.washingtonpost.com/climate-environment/2022/12/03/maine-lobster-whales-right/.

40. As quoted in Tony Davis, "Firebrand Ways," *High Country News*, December 28, 2009, https://www.hcn.org/issues/41.22/firebrand-ways.

41. Baier, *Inside the Equal Access*, 423–442, 289.

42. Lemann, "No People Allowed"; and Mitch Tobin, "Rancher Wins Big in Libel Suit Against Enviros," *High Country News*, February 21, 2005, https://www.hcn.org/issues/292/15300.

43. Martin, "In the Shadow of the Wolf," 88.

44. Alisa Opar, "Unprecedented Conservation Efforts Keep Greater Sage-Grouse Off Endangered Species List," *Audubon Magazine*, September 12, 2015, https://www.audubon.org/news/unprecedented-conservation-efforts-keep-greater-sage-grouse-endangered-species. Although these agreements were hopeful, it seems that backtracking by the Trump administration undid much of their promise. Andy McGlashen, "Frustration Abounds Five Years After the Historic Sage-Grouse Agreement," *Audubon Magazine*, September 2020, https://www.audubon.org/news/frustration-abounds-five-years-after-historic-sage-grouse-agreement.

45. Jedediah Britton-Purdy, "Environmentalism Was Once a Social-Justice Movement," *Atlantic*, December 7, 2016, https://www.theatlantic.com/science/archive/2016/12/how-the-environmental-movement-can-recover-its-soul/509831/.

46. Ted Williams, "Extreme Green," *High Country News*, May 31, 2011, https://www.hcn.org/wotr/extreme-green.

47. Annette McGivney, "Moses or Menace?," *Backpacker Magazine*, February 2003, 50.

48. Karl Mathiesen, "Protecting Nature, Destroying Lives," *Politico*, March 9, 2023, https://www.politico.eu/article/johan-vollenbroek-netherlands-nitrogen-pollution-climate-change-farming/; and Camille Gijs, "Police Fire on Dutch Farmers Protesting Environmental Rules," *Politico*, July 6, 2022, https://www.politico.eu/article/police-fire-dutch-farmer-protest-nitrogen-emission-cut/.

49. Nils-Gerrit Wunsch, "Leading Countries Worldwide by Value of Agricultural Products Exported in 2020," *Statista*, September 12, 2022, https://www.statista.com/statistics/1332329/leading-countries-worldwide-by-value-of-agricultural-products-exported/.

50. Erik Stokstad, "Nitrogen Crisis from Jam-Packed Livestock Operations Has 'Paralyzed' Dutch Economy," *Science*, December 4, 2019, https://www.science.org/content/article/nitrogen-crisis-jam-packed-livestock-operations-has-paralyzed-dutch-economy/; Isabeau van Halm, "The Dutch Nitrogen Crisis Shows What Happens When Policymakers Fail to Step Up," *Energy Monitor*, August 16, 2020, https://www.energymonitor.ai/policy/the-dutch-nitrogen-crisis-shows-what-happens-when-policymakers-fail-to-step-up/; and Wim de Vries, Han van Dobben, and Wieger Wamelink, "Kritische Depositiewaarden Zijn Bruitkbaar Voor Landelijk Beleid," *Nature Today*, September 5, 2022, https://www.naturetoday.com/intl/nl/nature-reports/message/?msg=29683.

51. Mathiesen, "Protecting Nature, Destroying Lives"; Nick Ottens, "The Netherlands' Farm Crisis, Explained," *Atlantic Sentinel*, June 12, 2022, https://atlanticsentinel.com/2022/06/the-netherlands-farm-crisis-explained/; and Hans Wetzels, "How a Nitrogen Crisis Turned Dutch Farmers to Rage," *Arc 2020*, November 24, 2019, https://www.arc2020.eu/nitrogen-crisis-dutch-farmers-rage/.

52. Mathiesen, "Protecting Nature, Destroying Lives."

53. Michael Durand-Wood, "Are Traffic Engineers Sociopaths?," *Strong Towns*, September 17, 2021, https://www.strongtowns.org/journal/2021/9/16/are-traffic-engineers-sociopaths.

54. James C. Scott, *Seeing Like a State: How Certain Schemes to Improve the Human Condition Have Failed* (New Haven, CT: Yale University Press, 1998), 193–222.

55. Jon Henley, "Rural Populist Party Emerges as Big Winner in Dutch Elections," *Guardian*, March 16, 2023, https://www.theguardian.com/world/2023/mar/16/rural-populist-party-farmer-citizen-movement-big-winner-dutch-elections.

56. My point, of course, is not that environmental law per se leads to populism, only that technocratic policies tend to evoke a populist reaction. As I will cover in the next chapter, environmental laws can be designed so as to foster participation and create opportunities for collective action. The outcome need not be polarized antagonism.

57. Tania Rabesandratana, "Scientists Urge European Parliament to Vote for Nature Restoration Law," *Science*, July 10, 2023, https://www.science.org/content/article/scientists-urge-european-parliament-vote-nature-restoration-law; and Guy Pe'er, Jana Kachler, Irina Herzon, Daniel Hering, Anni Arponen, Laura Bosco, et al., "Scientists Support the EU's Green Deal and Reject the Unjustified Argumentation Against the Sustainable Use Regulation and the Nature Restoration Law," *Zenodo*, July 9, 2023, https://doi.org/10.5281/zenodo.8033783.

58. Mathiesen, "Protecting Nature, Destroying Lives."

59. "The Maasai Taking Their Livestock into the Ngorongoro Caldera for Grazing" by Harvey Barrison, *Wikimedia*, May 30, 2012, https://commons.wikimedia.org/wiki/File:Ngorongoro_2012_05_30_2354_%287500935454%29.jpg, used under Creative Commons Attribution-Share Alike 2.0 Generic license, https://creativecommons.org/licenses/by-sa/2.0/deed.en.

60. Mark Dowie, *Conservation Refugees: The Hundred-Year Conflict Between Global Conservation and Native Peoples* (Cambridge, MA: MIT Press, 2009), xv–xvii.

61. Dowie, *Conservation Refugees*, xv–xvii.

62. Karl Jacoby, *Crimes Against Nature* (Berkeley: University of California Press, 2014), 11–78.

63. Tony Briscoe, "Conservationists See Rare Nature Sanctuary. Black Farmers See a Legacy Bought Out from Under Them," *ProPublica*, October 14, 2021, https://www.propublica.org/article/conservationists-see-rare-nature-sanctuaries-black-farmers-see-a-legacy-bought-out-from-under-them.

64. Catrin Einhorn, "Ecuador Strikes a Landmark Deal to Protect the Galápagos, and Save Cash," *New York Times*, May 9, 2023, https://www.nytimes.com/2023/05/09/climate/galapagos-ecuador-debt-nature.html.

65. Prakash Kashwan, "Inequality, Democracy, and the Environment: A Cross-National Analysis," *Ecological Economics* 131 (2017): 139–151.

66. Dowie, *Conservation Refugees*, 33; and Jacoby, *Crimes Against Nature*, 48–80.

67. Jonathan Watts, "Maasai Herders Driven Off Land to Make Way for Luxury Safaris, Report Says," *Guardian*, May 10, 2018, https://www.theguardian.com/environment/2018/may/10/maasai-herders-driven-off-land-to-make-way-for-luxury-safaris-report-says.

68. Jacoby, *Crimes Against Nature*, 39.

69. Nordhaus, "Two Visions Collide."

70. Watts, "Maasai Herders"; Katie McQue, "Tanzania's Maasai Appeal to West to Stop Eviction for Conservation Plans," *Guardian*, April 22, 2022, https://www.theguardian.com/global-development/2022/apr/22/tanzania-maasai-appeal-to-west-stop-evictions-due-to-conservation-plans; and Stephanie McCrummen, "This Will Finish Us," *Atlantic*, April 8, 2024, https://www.theatlantic.com/magazine/archive/2024/05/maasai-tribe-tanzania-forced-land-evictions-serengeti/677835/.

71. As quoted in Marina Lopes, "Otters Are Taking Over Singapore," *Washington Post*, October 22, 2021, https://www.washingtonpost.com/world/asia_pacific/singapore-otters-wildlife/2021/10/22/9e85c3ac-2afd-11ec-b17d-985c186de338_story.html/.

72. "Wild Otter Singapore" by Brian Evans, *flickr*, June 11, 2014, https://www.flickr.com/photos/66358983@N07/14395057284, used under Creative Commons Attribution-NoDerivs 2.0 Generic, https://creativecommons.org/licenses/by-nd/2.0/.

73. Lopes, "Otters Are Taking Over"; Rachel Tan, "Singapore's Adorable Otters, Explained," *Kopi*, April 30, 2021, https://thekopi.co/2021/04/30/from-love-stories-to-war-singapores-otters-are-surprisingly-human/; and Navene Elangova, "Visitor to Singapore Botanic Gardens 'Bitten 26 Times' by Otters; Such Incidents Are Rare, Says Gardens," *Today*, December 8, 2021, https://www.todayonline.com/singapore/visitor-singapore-botanic-gardens-bitten-26-times-otters-such-incidents-rare-says-gardens-176579.

74. "Death Loss in U.S. Cattle and Calves Due to Predator and Nonpredator Causes, 2015," *United States Department of Agriculture*, December 2017, https://www.aphis.usda.gov/animal_health/nahms/general/downloads/cattle_calves_deathloss_2015.pdf.

75. Martin, "In the Shadow of the Wolf," 51–90.

76. Alexander Zahl-Thanem, Rob J. F. Burton, Arild Blekesaune, Marit S. Haugen, and Katrina Rønningen, "The Impact of Wolves on Psychological Distress Among Farmers in Norway," *Journal of Rural Studies* 78 (2020): 1–11.

77. Rachel Nuwer, "The Problem with India's Man-Eating Tigers," *BBC Future*, November 21, 2019, https://www.bbc.com/future/article/20191120-the-problem-of-indias-man-eating-tigers; and Niranjana Rajalakshmi, "India's 'Man-Eating' Tigers Entangled in a Blame Game," *Undark*, May 16, 2022, https://undark.org/2022/05/16/indias-man-eating-tigers-entangled-in-a-blame-game/.

78. Dax, *Grizzly West*, 86.

79. Martin, "In the Shadow of the Wolf," 35.

80. Peter S. Alagona, *Accidental Ecosystems: People and Wildlife in American Cities* (Berkeley: University of California Press, 2022).

81. Tan, "Singapore's Adorable Otters."

82. Nordhaus, "Two Visions Collide."

83. Monbiot, "Lab-Grown Food Will Soon Destroy Farming"; and PoliticsJOE, "George Monbiot Breaks Down Why We Need to Destroy Farming," YouTube, 32:27, March 31, 2023, https://www.youtube.com/watch?v=534E9_7Dmg8.

84. George Monbiot and Tom Martin, "George Monbiot and Tom Martin on the End—or Rebirth—of Farming," *Prospect Magazine*, September 8, 2022, https://www.prospectmagazine.co.uk/politics/60049/george-monbiot-and-tom-martin-on-the-endor-rebirthof-farming; and Philippa Nuttall, "George Monbiot: 'Agriculture Is Arguably the Most Destructive Industry on

Earth,'" *The New Statesman*, May 13, 2022, https://www.newstatesman.com/encounter/2022/05/george-monbiot-agriculture-is-arguably-the-most-destructive-industry-on-earth.

85. Emily Moon, "Democrats Are Not Trying to Take Away Your Hamburger," *Pacific Standard*, March 1, 2019, https://psmag.com/news/democrats-are-not-trying-to-take-away-your-hamburgers; and John Henley, Sam Jones, Angela Giuffrida, and Philip Oltermann, "Greens v 'Beefatarians': Europeans Go to War over Their Dinner," *Guardian*, January 21, 2022, https://www.theguardian.com/environment/2022/jan/21/the-greens-want-to-take-our-meat-away-europeans-go-to-war-over-their-dinner.

86. Guy Pe'er (@GuyPeer3), "Quite a useful article on the polarized debate around meat consumption in Europe. Pity that it misses to the voice of science, indicating that meat consumption isn't an issue between greens and beefatarians, but rather, between humans and Earth," Twitter (now X), January 23, 2022, 1:34 p.m., https://twitter.com/GuyPeer3/status/1486075085421219842.

87. Pablo Manzano, Guillermo Pardo, Moustapha A. Itani, and Agustín del Prado, "Underrated Past Herbivore Density Could Lead to Disoriented Sustainability Policies," *npj Biodiversity* 2, no. 2 (2023), https://doi.org/10.1038/s44185-022-00005-z.

88. Pablo Manzano, Agustín del Prado, and Guillermo Pardo, "Comparable GHG Emissions from Animals in Wildlife and Livestock-Dominated Savannas," *npj Climate and Atmospheric Science* 6, no. 27 (2023), https://doi.org/10.1038/s41612-023-00349-8.

89. Dan Blaustein-Rejto, Alex Smith, and Emma Kovak, "The Clean Cow," *Breakthrough Institute*, October 21, 2021, https://thebreakthrough.org/issues/food-agriculture-environment/the-clean-cow.

90. Ryan Katz-Rosene, Flaminia Ortenzi, Graham A. McAuliffe, and Ty Beal, "Levelling Foods for Priority Micronutrient Value Can Provide More Meaningful Environmental Footprint Comparisons," *Communications Earth & Environment* 4, no. 1 (2023): 287.

91. Jim Robbins, "Where the Bison Could Roam," *New York Times*, January 10, 2023, https://www.nytimes.com/2023/01/10/science/bison-prairie-grassland.html; and Kyle E. Harms, "Bison Outperform Cattle at Restoring Their Home on the Range," *Proceedings of the National Academy of Sciences* 119, no. 40 (2022): e221363229.

92. Fred Pearce, "Sparing vs. Sharing: The Great Debate over How to Protect Nature," *Yale E360*, December 3, 2018, https://e360.yale.edu/features/sparing-vs-sharing-the-great-debate-over-how-to-protect-nature.

93. Anna Jones, "Apocalypse Cow: The Moment Monbiot Went Too Far," *Eat Farm Now*, January 24, 2020, https://eatfarmnow.com/2020/01/24/apocalypse-cow-the-moment-monbiot-went-too-far/.

94. IPBES, *Summary for Policymakers of the Methodological Assessment Report on the Diverse Values and Valuation of Nature of the Intergovernmental Science-Policy Platform on Biodiversity and Ecosystem Services*, ed. Unai Pascual et al. (Bonn, Germany: IPBES Secretariat, 2022), 31.

95. As quoted in Jacoby, *Crimes Against Nature*, 13.

96. Charles C. Mann, "The Book That Incited a Worldwide Fear of Overpopulation," *Smithsonian Magazine*, January 2018, https://www.smithsonianmag.com/innovation/book-incited-worldwide-fear-overpopulation-180967499/.

97. Naomi Klein, "Capitalism vs. the Climate," *The Nation*, November 9, 2022, https://www.thenation.com/article/archive/capitalism-vs-climate/.

98. IPBES, *Summary for Policymakers*, 37.

99. IPBES, *Methodological Assessment Report on the Diverse Values and Valuation of Nature of the Intergovernmental Science-Policy Platform on Biodiversity and Ecosystem Services*, ed. Patricia Balvanera et al. (Bonn, Germany: IPBES Secretariat, 2022), 347–440; and Roger Pielke Jr., "Does Climate Research Have an Anti-Growth Problem?," *The Honest Broker*, August 11, 2023, https://rogerpielkejr.substack.com/p/does-climate-research-have-an-anti?publication_id=119454&post_id=135934498&isFreemail=true.

100. Klein, "Capitalism vs. the Climate."

101. Carlo Invernizzi Accetti, "Repoliticizing Environmentalism: Beyond Technocracy and Populism," *Critical Review* 33, no. 1 (2021): 47–73.

102. Hillenkamp, "Die Panik vor den Rettern."

103. Rich, *Getting to Green.*

104. Peter Voosen, "Conservation: Myth-Busting Scientist Pushes Green's Reliance on 'Horror Stories,'" *EENews*, March 4, 2012, https://climateaccess.org/system/files/Voosen_Myth-Busting%20Scientist.pdf.

105. "Bridging the Rural–Urban Divide on Environmental Issues," *American University News*, April 21, 2020, https://www.american.edu/spa/news/bridging-the-rural-urban-divide-on-environmental-issues.cfm.

106. Brian Eckhouse and Naureen S. Malik, "Solar Boom Defies Politics in Spread to Red Midwest States," *Bloomberg Law*, February 7, 2023, https://news.bloomberglaw.com/environment-and-energy/solar-boom-defies-politics-in-spreading-to-red-corner-of-midwest.

107. Maddie Davis, "Gov. Gianforte Appealed Judge's APR Bison Decision," *Sidney Herald*, February 10, 2023, https://www.sidneyherald.com/news/agriculture/gov-gianforte-appealed-judges-apr-bison-grazing-decision/article_746d3c3e-8b7e-11ed-a7a8-a34c0b618e0d.html; Hunter Herbaugh, "Montana Stockgrowers Association Raising Funds to Fight Grazing Decision," May 10, 2023, https://www.terrytribune.com/news/montana-stockgrowers-association-raising-funds-to-fight-grazing-allotment-decision/article_10b6ab56-ef44-11ed-b44f-47107d91ad65.html; and Jake Bullinger, "Montana Refuge Divides Tribes and Ranchers," *High Country News*, April 28, 2017, https://www.hcn.org/issues/49.9/montana-prairie-refuge-divides-natives-and-ranchers.

108. Nordhaus, "Two Visions Collide"; and Bullinger, "Montana Refuge Divides Tribes."

109. Anne Herbst and Aaron Adelson, "'A Very Highly Emotionally Charged Issue': How Wolves Have Divided Colorado," *9 News*, May 19, 2023, https://www.9news.com/article/life/animals/colorado-wolves-reintroduction-divide-prop-114/73-b65b6124-4a42-49f5-bd2f-a8f612729358.

110. Sarah Gibbens, "Activists Fear a New Threat to Biodiversity—Renewable Energy," *National Geographic*, May 26, 2022, https://www.nationalgeographic.com/environment/article/activists-fear-biodiversity-threat-from-renewable-energy; and Jack Forrest, "Greens Ask Feds to Protect Rare Nevada Butterfly," *Greenwire*, October 10, 2022, https://subscriber.politicopro.com/article/eenews/2022/08/10/greens-ask-feds-to-protect-rare-nevada-butterfly-00050797.

111. Colin J. Whitfield, Helen M. Baulch, Kwok P. Chun, and Cherie J. Westbrook, "Beaver-Mediated Methane Emissions: The Effects of Population Growth in Eurasia and the Americas," *AMBIO* 44 (2015): 7–15; and Environmental Protection Agency, *Inventory of U.S. Greenhouse Gas Emissions and Sinks*, Washington, DC, 2017, https://www.epa.gov/sites/default/files/2017-02/documents/2017_complete_report.pdf.

CHAPTER 6

1. "Botswana Vows to Send 20,000 Elephants to Germany in Hunting Row," *Euractiv*, April 3, 2024, https://www.euractiv.com/section/all/news/botswana-vows-to-send-elephants-to-germany-in-hunting-row/.

2. Cari Shane, "Living 'Bee Fences' Protect Farmers from Elephants, and Vice Versa," *Scientific American*, September 27, 2021, https://www.scientificamerican.com/article/living-bee-fences-protect-farmers-from-elephants-and-vice-versa/; Hannah Thomasy, "Beehive Fences Can Help Mitigate Human–Elephant Conflict," *Mongabay*, September 11, 2019, https://news.mongabay.com/2019/09/beehive-fences-can-help-mitigate-human-elephant-conflict/; Anne Pinto-Rodrigues, "A Taste of Honey: How Bees Mend Fences Between Farmers and Elephants," *Guardian*, August 12, 2020, https://www.theguardian.com/environment/2020/aug/12/a-taste-of-honey-how-bees-mend-fences-between-farmers-and-elephants-aoe; and Lucy King, Lydia Tiller, Emmanuel Mwambingu, Esther Serem, Hesron Nzumu, Gloria Mugo, et al., "Impact of Drought and Development on the Effectiveness of Beehive Fences as Elephant Deterrents over 9 Years in Kenya," *Conservation Science and Practice* 6, no. 11 (2024): e13242.

3. Thomasy, "Beehieve Fences"; Pinto-Rodrigues, "A Taste of Honey"; and King et al., "Impact of Drought and Development."

4. Michael Rosenzweig, *Win-Win Ecology: How the Earth's Species Can Survive in the Midst of the Human Enterprise* (New York: Oxford University Press, 2003), 7.

5. Rosenzweig, *Win-Win Ecology*.

6. Mike Shanahan, "How Fig Trees Could Revolutionise Reforestation and Tourism in Borneo," *China Dialogue*, January 17, 2024, https://chinadialogue.net/en/nature/borneo-fig-trees-reforestation-wildlife-tourism/.

7. "CD Aero Eliminates Use of nPB, Gains Production Capacity," *TURI*, September 2021, https://www.turi.org/content/download/13650/212560/file/CD+Aero+Eliminates+Use+of+nPB.pdf; and Ian Urbina, "As OSHA Emphasizes Safety, Long-Term Health Risks Fester," *New York Times*, March 30, 2013, https://www.nytimes.com/2013/03/31/us/osha-emphasizes-safety-health-risks-fester.html.

8. Ken Geiser, *Chemicals Without Harm: Policies for a Sustainable World* (Cambridge, MA: MIT Press, 2015), 1–58.

9. Geiser, *Chemicals Without Harm*; and Environmental Protection Agency, *Report to Congress on EPA's Capacity to Implement Certain Provisions of the Frank R. Lautenberg Chemical Safety for the 21st Century Act*, Washington DC, October 2022, https://www.epa.gov/system/files/documents/2022-10/October%202022%20Report%20to%20Congress%20on%20the%20EPAs%20Capacity%20to%20Implement%20Certain%20Provisions%20of%20the%20Frank%20R.%20Lautenberg%20Chemical%20Safety%20for%20the%2021st%20Century%20Act.pdf.

10. Silvio O. Funtowicz and Jerome R. Ravetz, "Uncertainty, Complexity, and Post-Normal Science," *Environmental Toxicology and Chemistry* 13, no. 12 (1994): 1881–1885.

11. Author creation. The idea to depict stakes and uncertainties of post-normal science as *x–y*-axes on a figure comes from Funtowicz and Ravetz, "Uncertainty, Complexity, and Post-Normal Science,"1882.

12. Oreskes and Conway, *Merchants of Doubt*.

13. Daniel Sarewitz, "Political Effectiveness in Science and Technology," in *Science in the Context of Application*, ed. Martin Carrier and Alfred Nordmann (New York: Springer, 2011), 308.

14. Rayner and Sarewitz, "Policy Making in the Post-Truth World."

15. Molly M. Jacobs, Rachel I. Massey, Heather Tenney, and Elizabeth Harriman, "Reducing the Use of Carcinogens: The Massachusetts Experience," *Review of Environmental Health* 29, no. 4 (2014): 319–340; and Environmental Protection Agency, "EPA Reported Significant Drop in Toxic Chemical Releases in 2019," *Inspectioneering*, January 14, 2021, https://inspectioneering.com/news/2021-01-14/9492/epa-reported-significant-drop-in-toxic-chemical-releases-in-2019.

16. TURI features what political scientist Charles Lindblom described as the foundation of good problem-solving: "to create . . . opportunities for collective action that did not exist before." It doesn't try to force a solution but rather enables the large-scale coordination of decision making to address a social problem. See Charles Lindblom, *Inquiry and Change: The Troubled Attempt to Understand and Shape Society* (New Haven, CT: Yale University Press, 1990), 244.

17. David Christian Rose, "Avoiding a Post-Truth World: Embracing Post-Normal Conservation," *Conservation and Society* 16, no. 4 (2018): 518–524.

18. Lisa Song, "Out of Balance," *ProPublica*, June 15, 2023, https://www.propublica.org/article/biodiversity-offsets-guinea-world-bank-group-chimpanzees-outbreak.

19. Song, "Out of Balance"; and Jennifer O'Mahony, "Bauxite Mining and Chinese Dam Push Guinea's Chimpanzees to the Brink," *Mongabay*, May 21, 2019, https://news.mongabay.com/2019/05/bauxite-mining-and-chinese-dam-push-guineas-chimpanzees-to-the-brink/.

20. Peter Kareiva, "New Conservation: Setting the Record Straight and Finding Common Ground," *Conservation Biology* 28, no. 3 (2014): 634–636.

21. Jacob P. Byl, "Perverse Incentives and Safe Harbors in the Endangered Species Act: Evidence from Timber Harvests Near Woodpeckers," *Ecological Economics* 157 (2019): 100–108; and Peter S. Alagona, *After the Grizzly: Endangered Species and the Politics of Place in California* (Berkeley: University of California Press, 2013), 192.

22. Alagona, *After the Grizzly*, 149–174

23. Louis Sahagún, "California's Mojave Desert Tortoises Move Toward Extinction. Why Saving Them is so Hard," *Los Angeles Times*, November 17, 2022, https://www.latimes.com/environment/story/2022-11-17/biologists-fear-desert-tortoise-is-headed-for-extinction.

24. Jonathan Allen, "Before Nevada Stand-Off, A Collison Between Ranchers and Tortoises," *Reuters*, May 30, 2014, https://www.reuters.com/article/usa-ranchers-nevada-tortoises-insight-pi-idINL1N0O12F920140530/.

25. "Poverty Rate by Country 2023," *World Population Review*, accessed August 16, 2023, https://worldpopulationreview.com/country-rankings/poverty-rate-by-country; "Guinea," *Transparency International*, accessed August 16, 2023, https://www.transparency.org/en/countries/guinea; and Rachel Chason and Chloe Sharrock, "On Frontier of New 'Gold Rush,' Quest for Coveted EV Metals Yields Misery," *Washington Post*, April 27, 2023, https://www.washingtonpost.com/world/interactive/2023/ev-battery-bauxite-guinea/.

26. Katie Devenish, Sébastien Desbureaux, Simon Willcock, and Julia P. G. Jones., "On Track to Achieve No Net Loss of Forest at Madagascar's Biggest Mine," *Nature Sustainability* 5 (2022): 498–508; and Sophus O. S. E. zu Ermgassen, Julie Baker, Richard A. Griffiths, Niels Strange, Matthew J. Struebig, and Joseph W. Bull, "The Ecological Outcomes of Biodiversity Offsets Under 'No Net Loss' Policies: A Global Review," *Conservation Letters* 12, no. 6 (2019): e12664.

27. Song, "Out of Balance"; and O'Mahony, "Bauxite Mining and Chinese Dam."

28. Martine Maron, "Is 'No Net Loss of Biodiversity' a Good Idea?," in *Effective Conservation Science: Data not Dogma*, ed. Peter Kareiva, Michelle Marvier, and Brian Silliman (New York: Oxford University Press, 2017), 141–146.

29. Song, "Out of Balance."

30. Cécile Bidaud, Kate Schreckenberg, Manolotsoa Rabeharison, Patrick Ranjatson, James Gibbons, and Julia P. G. Jones, "The Sweet and the Bitter: Intertwined Positive and Negative Social Impacts of a Biodiversity Offset," *Conservation and Society* 15, no. 1 (2017): 1–13.

31. Song, "Out of Balance."

32. Jasper R. de Vries, Eva van der Zee, Raoul Beunen, Rianne Kat, and Peter H. Feindt, "Trusting the People and the System. The Interrelation Between Interpersonal and Institutional Trust in Collective Action for Agri-Environmental Management," *Sustainability* 11, no. 24 (2019): 7022.

33. de Vries et al., "Trusting the People."

34. de Vries et al., "Trusting the People."

35. Nicolas Grondard, Erik Kleyheeg, Lars Hein, Lenny and G. J. Van Bussel, "Effects of Dutch Agri-Environmental Field Margins and Bird Plots on Cropland Birds," *Agriculture, Ecosystems and Environment* 349, no. 15 (2023): 108430.

36. Holly Firlein, "Continental Divides: How Wolf Conservation in the United States and Europe Impacts Rural Attitudes," *Ecological Law Quarterly* 45, no. 2 (2018): 327–352.

37. Robert B. Wielgus and Kaylie A. Peebles, "Effects of Wolf Mortality on Livestock Depredations," *PLoS One* 9, no. 12 (2014): e113505.

38. Jeff Vance Martin, "Peace in the Valley? Qualitative Insights on Collaborative Coexistence from the Wood River Wolf Project," *Conservation Science and Practice* 3, no. 3 (2021): e197; and Max G. Levy, "Science Can Keep Wolves from Killing Livestock," *Smithsonian Magazine*, December 11, 2020, https://www.smithsonianmag.com/science-nature/these-non-lethal-methods-encouraged-science-can-keep-wolves-killing-livestock-180976505/.

39. George Wuerthner, "Coexistence Between Wolves and Livestock Is a Delusion," *High Country News*, October 19, 2017, https://www.hcn.org/articles/opinion-coexistence-between-wolves-and-livestock-is-a-delusion.

40. Taylor Dotson and Henrique M. Pereira, "From Antagonist Conservation to Biodiversity Democracy in Rewilding," *One Earth* 5, no. 5 (2022): 466–469.

41. Jens Persson, Geir R. Rauset, and Guillaume Chapron, "Paying for an Endangered Predator Leads to Population Recovery," *Conservation Letters* 8, no. 5 (2015): 345–350.

42. Alix Soliman, "The Hidden Costs of Wolf Conservation," *High Country News*, May 1, 2025, https://www.hcn.org/issues/57-5/the-hidden-costs-of-wolf-conservation/.

43. de Vries et al., "Trusting the People."

44. de Vries et al., "Trusting the People."

45. Jason Bittel, "Why Cecil the Lion Was So Popular with People," *National Geographic*, July 30, 2015, https://www.nationalgeographic.com/culture/article/150730-cecil-lion-africa-hunting-science-animals.

46. "How the Internet Descended on the Man Who Killed Cecil the Lion," *BBC News*, July 29, 2015, https://www.bbc.com/news/blogs-trending-33694075; "Cecil the Lion Petition Reaches White House," *Sky News*, July 30, 2015, https://news.sky.com/story/cecil-the-lion-petition-reaches-white-house-10350847; and Abby Reimer, "For Many in Zimbabwe, Cecil the Lion Tells Only Part of the Story," *CNN*, August 3, 2015, https://www.cnn.com/2015/08/03/africa/zimbabwe-other-problems-cecil-the-lion/index.html.

47. Richard Conniff, "A Trophy Hunt That's Good for Rhinos," *New York Times*, January 20, 2014, https://www.nytimes.com/2014/01/21/opinion/a-trophy-hunt-thats-good-for-rhinos.html; and Brian John Huntley, *Strategic Opportunism: What Works in Africa* (Cham, Switzerland: Springer, 2023), 93–120.

48. Conniff, "A Trophy Hunt"; and Huntley, *Strategic Opportunism.*

49. Hannah Hoag, "In Canada's Boreal Forest, a New National Park Faces the Wrongs of the Past—and Guards Our Climate Future," *Audubon Magazine*, Fall 2019, https://www.audubon.org/magazine/fall-2019/in-canadas-boreal-forest-new-national-park-faces.

50. Agnes Kalibata, "Indigenous Peoples Are the Best Stewards of our Environment—the Rest of Us Pale in Comparison," *The Independent*, November 5, 2021, https://www.independent.co.uk/climate-change/opinion/cop26-farmers-food-indigenous-peoples-climate-crisis-b1951423.html?r=9522.

51. See Allyn MacLean Stearman, "'Only Slaves Climb Trees': Revisiting the Myth of the Ecologically Noble Savage in Amazonia," *Human Nature* 5, no. 4 (1994): 339–357.

52. Chelsey Geralda Armstrong, Jesse Miller, Alex McAlvay, Patrick Morgan Ritchie, Dana Lepofsky, "Historical Indigenous Land-Use Explains Plant Functional Trait Diversity," *Ecology & Society* 26, no. 2 (2021): 6; Milton M. R. Freeman, "The Nature and Utility of Traditional Ecological Knowledge," *Northern Perspectives* 20, no. 1 (1992): 9–12; and Aarti Betigeri, "How Australia's Indigenous Experts Could Help Deal with Devastating Wildfires," *Time*, January 14, 2020, https://time.com/5764521/australia-bushfires-indigenous-fire-practices/.

53. Shepard Krech III, *The Ecological Indian: Myth and History* (New York: W. W. Norton & Company, 1999), 123–150.

54. Morten Erik Allentoft, Rasmus Heller, Charlotte L. Oskam, Eline D. Lorenzen, Marie L. Hale, M. Thomas P. Gilbert, et al., "Extinct New Zealand Megafauna Were Not in Decline Before Human Colonization," *Proceedings of the National Academy of Sciences* 111, no. 13 (2014): 4922–4927.

55. Fikret Berkes, *Sacred Ecology* (New York: Routledge, 2008), 117–139.

56. Stearman, "'Only Slaves Climb Trees,'" 3492–350.

57. Berkes, *Sacred Ecology*, 241–245. One indigenous leader, Nicanor Gonzalez, put the matter bluntly: "We aren't nature lovers. At no time have indigenous groups included the concepts of conservation and ecology in their traditional vocabulary," as quoted in Stearman, "'Only Slaves Climb Trees,'" 352.

58. Cronon, "The Trouble with Wilderness."

59. Jon Mikkel Eira, "How Blood Pancakes and Boiled Meat Keep Traditions Alive," *Vice*, September 21, 2017, https://www.vice.com/en/article/evpxn4/how-blood-pancakes-and-boiled-meat-keep-traditions-alive.

60. Michelle Cyca, "Vancouver's New Mega-Development Is Big, Ambitious and Undeniably Indigenous," *Maclean's*, March 11, 2024, https://macleans.ca/society/sen%CC%93a%E1%B8%B5w-vancouver.

61. As quoted in Manvir Singh, "It's Time to Rethink the Idea of the 'Indigenous,'" *New Yorker*, February 20, 2023, https://www.newyorker.com/magazine/2023/02/27/its-time-to-rethink-the-idea-of-the-indigenous. Earlier examples included a movement in the Brazilian parliament to reduce indigenous lands after finding out the Kayapo were selling gold and lumber from their reserve. Stearman, "'Only Slaves Climb Trees,'" 351–352.

62. Cronon, "The Trouble with Wilderness," 21.

63. Richard Conniff, "In Namibia, a Lauded Community Conservation Model Is at Risk," *YaleEnvironment360*, June 22, 2023, https://e360.yale.edu/features/namibia-wildlife-conservation-corruption; Huntley, *Strategic Opportunism*; and Frowin Becker, "Reframing Trophy Hunting's Socio-Economic Benefits in Namibia," *Mongabay*, April 28, 2022, https://news.mongabay.com/2022/04/reframing-trophy-huntings-socio-economic-benefits-in-namibia-commentary/.

64. Carolyn K. Lesorogol, *Conservation and Community in Kenya: Milking the Elephant* (Lanham, MA: Lexington Books, 2022).

65. Patrick Barkham, "From Window Boxes to Farms, UK's Rewilding Gurus Bring Out their Bibles," *Guardian*, June 13, 2023, https://www.theguardian.com/environment/2023/jun/13/from-window-boxes-to-farms-uks-rewilding-gurus-bring-out-their-bible; and Elizabeth Fitt, "From Debt to Diversity: A Journey of Rewilding, Carbon Capture and Hope," *Mongabay*, August 9, 2023, https://news.mongabay.com/2023/08/from-debt-to-diversity-a-journey-of-rewilding-carbon-capture-and-hope/.

66. Fitt, "From Debt to Diversity"; and Barkham, "From Window Boxes to Farms."

67. George Monbiot, Twitter (now X), August 25, 2018, 1:20 a.m., https://twitter.com/GeorgeMonbiot/status/1033252883532136449.

68. Phoebe Weston, "Knepp Estate: Why the King and Queen of Rewilding Are Farming Again After 20 Years," *Guardian*, May 26, 2022, https://www.theguardian.com/environment/2022/may/26/knepp-estate-why-the-king-and-queen-of-rewilding-are-farming-again-after-20-years-aoe.

69. Matthew Hayek and Rachael D. Garrett, "Nationwide Shift to Grass-Fed Beef Requires Larger Cattle Population," *Environmental Research Letters* 13, no. 8 (2018): 084005; and Randall D. Jackson, "Grazed Perennial Grasslands Can Match Current Beef Production While Contributing to Climate Mitigation and Adaptation," *Agricultural & Environmental Letters* 7, no. 1 (2022): e20059.

70. Pablo Tittonell, Gervasio Pineiro, Lucas A. Garibaldi, Santiago Dogliotti, Han Olff, and Esteban G. Jobbagy, "Agroecology in Large Scale Farming—A Research Agenda," *Frontiers in Sustainable Food Systems* 18 (2020), https://doi.org/10.3389/fsufs.2020.584605.

71. Manjuantah G., Sandeep Hanchanale, and Veena Srinivasan, "Is Agroecological Farming a Solution for Rainfed Degraded Land?," *Mongabay*, October 4, 2022, https://india.mongabay.com/2022/10/commentary-is-agroecological-farming-a-solution-for-rainfed-degraded-land/.

72. Ahmed Hassanali, Hans Herren, Zeyaur R. Khan, John A. Pickett, and Christine M. Woodcock, "Integrated Pest Management: The Push–Pull Approach for Controlling Insect Pests and Weeds of Cereals, and its Potential for Other Agricultural Systems Including Animal Husbandry," *Philosophical Transactions of the Royal Society B* 363, no. 1491 (2008): 611–621.

73. Liz Carlisle and Niki Mazaroli, "The Return of Silvopasture," *Aeon*, March 30, 2023, https://aeon.co/essays/heres-to-reviving-the-ancient-practice-of-silvopasture.

74. Nassib Mugwanya, "After Agroecology," *The Breakthrough Journal*, February 4, 2019, https://thebreakthrough.org/journal/no-10-winter-2019/after-agroecology.

75. Laljeet Sangha, "Revisiting the Impacts of the Green Revolution in India," *Institute for Policy and Governance*, accessed September 19, 2023, https://ipg.vt.edu/DirectorsCorner/re--reflections-and-explorations/Reflections101520.html; and Daisy A. John and Giridhara R. Babu, "Lesson from the Aftermaths of Green Revolution on Food Systems and Health," *Frontiers in Sustainable Food Systems* 5 (2023), https://doi.org/10.3389/fsufs.2021.644559.

76. Tristan Kennedy, "The Uncomfortable Truth About 'Rewilding,'" *Vice*, November 4, 2020, https://www.vice.com/en/article/dy88eq/rewilding-climate-crisis-solution.

77. "Seadrift Operations," *Dow*, accessed August 22, 2023, https://corporate.dow.com/en-us/locations/seadrift.html; Dow, "Seadrift Constructed Wetlands: An Engineering Natural Technology Project," accessed May 29, 2025, https://corporate.dow.com/documents/science-sustainability/066-00203-01-01-seadrift-constructed-wetlands-case-study.pdf; and Johnathan L. DiMuro, France M. Guertin, Rich K. Helling, Jessica L. Perkins, and Scanlon Romer, "A Financial and Environmental Analysis of Constructed Wetlands for Industrial Wastewater Treatment," *Journal of Industrial Ecology* 18, no. 5 (2014): 631–640.

78. "Seadrift Operations"; and "Seadrift Constructed Wetlands."

79. Pelayo Menéndez, Iñigo J. Losada, Saul Torres-Ortega, Siddharth Narayan, and Michael W. Beck, "The Global Flood Protection Benefits of Mangroves," *Scientific Reports* 10, no. 4404 (2020), https://doi.org/10.1038/s41598-020-61136-6; Julie Rouet-Leduc, "Effects of Large Herbivores on Fire Regimes and Wildfire Mitigation," *Journal of Applied Ecology* 58, no. 12 (2021): 2690–2701; and Carrie Arnold, "Wildfire of Varying Intensity Can Be Good for Biodiversity," *Quanta Magazine*, November 29, 2021, https://www.quantamagazine.org/wildfires-of-varying-intensity-can-be-good-for-biodiversity-20211129/.

80. Timm Kroeger, Francisco J. Escobedo, José L. Hernandez, Sebastián Varela, Sonia Delphin, Jonathan R. B. Fisher, et al., "Reforestation as a Novel Abatement and Compliance Measure for Ground-Level Ozone," *Proceedings of the National Academy of Sciences* 111, no. 40 (2014): E4204–E4213.

81. David Tilman and John A. Downing, "Biodiversity and Stability in Grasslands," *Nature* 367 (1994): 363–365; and Joseph E. Fargione and David Tilman, "Diversity Decreases Invasion via Both Sampling and Complementary Effects," *Ecology Letters* 8 (2005): 604–611.

82. Ken Thompson, *Do We Need Pandas?* (Totnes, UK: Green Books, 2010), 109. See also Peter Kareiva and Michelle Marvier, "Conserving Biodiversity Coldspots: Recent Calls to Direct

Conservation Funding to the World's Biodiversity Hotspots May Be Bad Investment Advice," *American Scientist* 91, no. 4 (2003): 344–351.

83. Forest Isbell, Vincent Calcagno, Andy Hector, John Connolly, W. Stanley Harpole, Peter B. Reich, et al., "High Plant Diversity Is Needed to Maintain Ecosystem Services," *Nature* 477 (2011): 199–202.

84. Paul J. Ferraro, "Are Payments for Ecosystem Services Benefiting Ecosystems and People?," in *Effective Conservation Science: Data not Dogma*, ed. Peter Kareiva, Michelle Marvier, and Brian Silliman (New York: Oxford University Press, 2017), 159–172; and Alisher Mirzabaev and David Wuepper, "Economics of Ecosystem Restoration," *Annual Review of Resource Economics* 15, no. 4 (2023): 4.1–4.22.

85. Marris, *Rambunctious Garden*, 166.

86. Jennifer L. Molnar, "Corporations Valuing Nature: It's Not All About the Win-Wins," in *Effective Conservation Science: Data not Dogma*, ed. Peter Kareiva, Michelle Marvier, and Brian Silliman (New York: Oxford University Press, 2017), 167–172.

87. Brett Bryan, "Nature Is Collapsing—What Must We Do?," *Disruptr*, September 13, 2019, https://disruptr.deakin.edu.au/environment/nature-is-collapsing-what-must-we-do/.

88. J. R. Reilly, D. R. Artz, D. Biddinger, K. Bobiwash, N. K. Boyle, C. Brittain, et al., "Crop Production in the USA Is Frequently Limited by a Lack of Pollinators," *Proceedings of the Royal Society B* 287 (2020): 20200922. The proportion of pollination done by domestic honeybees and wild insects can vary dramatically by crop, but *on average* they make an equal contribution, at least according to existing agricultural studies. James Reilly, Ignasi Bartomeus, Dylan Simpson, Alfonso Allen-Perkins, Lucas Garibaldi, and Rachael Winfree, "Wild Insects and Honey Bees Are Equally Important to Crop Yields in a Global Analysis," *Global Ecology and Biogeography* 33, no. 7 (2024): e13843.

89. "Bishan Park, aerial view" by Atelierdreiseitl, *Wikimedia*, June 5, 2012, https://commons.wikimedia.org/wiki/File:Singapore_Bishan_Park_Aerial.jpg, used under Creative Commons Attribution-Share Alike 3.0 Unported license, https://creativecommons.org/licenses/by-sa/3.0/deed.en.

90. Kuei-Hsien Liao, "The Socio-Ecological Practice of Building Blue-Green Infrastructure in High-Density Cities: What Does the ABC Waters Program in Singapore Tell Us?," *Socio-Ecological Practice Research* 1, no. 1 (2019): 67–81.

91. Erik Schofield, "How Bishan Park Became 'The Central Park' of Singapore," *Land8*, April 20, 2015, https://land8.com/how-bishan-park-became-the-central-park-of-singapore/.

92. Cronon, "The Trouble with Wilderness."

93. Zuzanna M. Rosin, Tomas Pärt, Matthew Low, Dorota Kotowska, Marcin Tobółka, Paweł Szymański, et al., "Village Modernization May Contribute More to Farmland Bird Declines Than Agricultural Intensification," *Conservation Letters* 14, no. 6 (2021): e12843.

94. Robert A. Francis and Jamie Lorimer, "Urban Reconciliation Ecology: The Potential of Living Roofs and Walls," *Journal of Environmental Management* 92, no. 6 (2011): 1429–1437;

Phoebe Weston, "Buzz Stops: Bus Shelter Roofs Turned into Gardens for Bees and Butterflies," *Guardian*, September 24, 2022, https://www.theguardian.com/environment/2022/sep/24/bus-shelter-roofs-turned-into-gardens-for-bees-butterflies-aoe; and Alessandro Filazzola, Namrata Shrestha, and J. Scott MacIvor, "The Contribution of Constructed Green Infrastructure to Urban Biodiversity: A Synthesis and Meta-Analysis," *Journal of Applied Ecology* 56, no. 9 (2019): 2131–2143.

95. Nathalie Seddon, Alexandre Chausson, Pam Berry, Cécile A. J. Girardin, Alison Smith, and Beth Turner, "Understanding the Value and Limits of Nature-Based Solutions to Climate Change and Other Global Challenges," *Philosophical Transactions of the Royal Society B* 375, no. 1794 (2020): 20190120.

96. Belmont Lay, "Otter Makes Bishan Park Its New Home, Causes Everybody to Say 'Awww,'" *Mothership*, May 10, 2014, https://mothership.sg/2014/05/otter-makes-bishan-park-its-new-home-causes-everybody-to-say-awww/.

97. Amanda J. Sellmer and Barbara S. Goto, "Urban Wildlife Corridors: Building Bridges for Wildlife and People," *Frontiers in Sustainable Cities* 4 (2022), https://doi.org/10.3389/frsc.2022.954089; and "Wildlife Crossings," *National Geographic*, accessed September 8, 2023, https://education.nationalgeographic.org/resource/wildlife-crossings/11th-grade/.

98. Liao, "The Socio-Ecological Practice."

99. Philip Wallach, "The Star Wars Theory of History," *R Street*, February 16, 2019, https://www.rstreet.org/commentary/the-star-wars-theory-of-history/.

100. Jon Henley, "Few Willing to Change Lifestyle to Save the Planet, Climate Survey Finds," *Guardian*, November 21, 2021, https://www.theguardian.com/environment/2021/nov/07/few-willing-to-change-lifestyle-climate-survey.

101. Rosenzweig, *Win-Win Ecology*, 181.

102. Christopher Lasch, *The Minimal Self: Psychic Survival in Troubled Times* (New York: W. W. Norton and Company, 1984), 17. Environmentalists sometimes forget that hunters and fisherman, people who directly experience the benefits of nature protections, are often some of the most dedicated conservationists.

103. This is why I haven't made use of the recently developed concept of "convivial conservation." There is considerable overlap between reconciliation biodiversity and the convivialists' vision of a conservation paradigm that includes people, celebrates "everyday nature," and is governed democratically, but they ultimately champion a vision of nature and society that is explicitly built upon values that are only widespread among highly educated, left-wing professionals. They call for degrowth and moving "beyond capitalism." It is possible that doing so might end up being a necessity, but as I'll continue to articulate in the following chapters, I think we will only find out through the incremental mutual adjustments that we make within the world as it exists now. And, for the time being, that means that conservation will have to compromise with the large portion of people who are neither pro-degrowth nor anti-capitalism. See Bram Büscher and Robert Fletcher, "Towards Convivial Conservation," *Conservation & Society* 17, no. 3 (2019): 283–296.

104. Meredith Root-Bernstein and Jens-Christian Svenning, "Prospects for Rewilding with Camelids," *Journal of Arid Environments* 130 (2016): 54–61.

105. Erica Berry, "The Long, Slow Trek to Get Americans to Eat Camel Meat," *NPR*, February 2, 2017, https://www.npr.org/sections/thesalt/2017/02/02/512701346/the-long-slow-trek-to-get-americans-to-eat-camel-meat.

106. Jaber Rahimi, Erwann Fillol, John Y. Mutua, Giuseppina Cinardi, Timothy P. Robinson, An M. O. Notenbaert, et al., "A Shift from Cattle to Camel and Goat Can Sustain Milk Production with Lower Inputs and Emissions in North Sub-Saharan Africa's Drylands," *Nature Food* 3, no. 523–531 (2022), https://doi.org/10.1038/s43016-022-00543-6; and Bijal Trivedi, "Kangaroos to the Rescue," *New Scientist* 200, no. 2687/2688 (2008): 48–50.

CHAPTER 7

1. Caroline Cox and Luke Hauser, "Ice Ivory to White Gold: Links Between the Illegal Ivory Trade and the Trade in Geocultural Artifacts," *Journal of International Wildlife Law & Policy* 26, no. 1 (2023): 22–46; and Christopher Hastings, "Mammoth Ivory Hunting in Siberia: Economic, Environmental and Paleontological Considerations," *Polar Research* 44, no. 10874 (2025), https://doi.org/10.33265/polar.v44.10874.

2. Hastings, "Mammoth Ivory Hunting."

3. Ewen Callaway, "Million-Year-Old Mammoth Genomes Shatter Record for Oldest Ancient DNA," *Nature* 590 (2021): 537–538.

4. Christine Lagorio-Chafkin, "How This Founder Embraced His Critics and Raised $75 Million for a Mammoth Moonshot," *Inc.*, May 30, 2022, https://www.inc.com/christine-lagorio-chafkin/colossal-woolly-mammoth-ben-lamm-what-i-know-podcast.html; and David Bloom, "Colossal Raises $200 Million from Dodgers Co-Owner, Hollywood Mega Producer," *Forbes*, January 17, 2025, https://www.forbes.com/sites/dbloom/2025/01/16/colossal-biosciences-raises-200-million-for-de-extinction-efforts/.

5. 60 Minutes, "Siberia's Pleistocene Park: Bringing Back Pieces of the Ice Age to Combat Climate Change," YouTube, December 8, 2019, 13:37, https://www.youtube.com/watch?v=nEzskUGJ_1I.

6. Jackson Ryan, "Thylacine De-Extinction: Why We Need to Talk About Resurrecting Species," *CNET*, August 19, 2022, https://www.cnet.com/science/biology/thylacine-de-extinction-why-we-need-to-talk-about-resurrecting-species/.

7. Laura DeFrancesco, "One Year, 1 Lab, 16 Spinouts," *Nature Biotechnology* 38 (2020): 9–13; Andy Lamey, "The De-Extinction of the Woolly Mammoth Is a Legal and Regulatory Nightmare," *The New Republic*, December 12, 2020, https://newrepublic.com/article/169528/woolly-mammoth-alaska-colossal-thiel; and Jillian McCall, "Why Radical Transparency Is Not a Luxury but a Prerequisite for Effective Engagement," *Colossal*,

accessed October 30, 2023, https://colossal.com/why-radical-transparency-is-not-a-luxury/.

8. Heather Hansman, "Re-Engineering the Colorado River to Save the Grand Canyon," *Scientific American*, February 1, 2019, https://www.scientificamerican.com/article/re-engineering-the-colorado-river-to-save-the-grand-canyon/.

9. Hansman, "Re-Engineering the Colorado River."

10. Charles Lindblom, "The Science of 'Muddling Through,'" *Public Administration Review* 19, no. 2 (1959): 79–88.

11. Joseph G. Morone and Edward J. Woodhouse, *Averting Catastrophe: Strategies for Regulating Risky Technologies* (Berkeley: University of California Press, 1986); David Collingridge, *The Management of Scale: Big Organizations, Big Decisions, Big Mistakes* (New York: Routledge, 1992); and Matthias Heymann, "Signs of Hubris: The Shaping of Wind Technology Styles in Germany, Denmark, and the United States, 1940–1990," *Technology and Culture* 39, no. 4 (1998): 641–670.

12. Joseph G. Morone and Edward J. Woodhouse, *The Demise of Nuclear Energy? Lessons for Democratic Control of Technology* (New Haven: Yale University Press, 1989); and Collingridge, *The Management of Scale.*

13. Morone and Woodhouse, *Averting Catastrophe*; and Heymann, "Signs of Hubris."

14. Morone and Woodhouse, *Averting Catastrophe*; and Collingridge, *The Management of Scale.*

15. Carl J. Walters and C. S. Holling, "Large-Scale Management Experiments and Learning by Doing," *Ecology* 71, no. 6 (1990): 2060–2068.

16. Ronald M. Thom, "Adaptively Addressing Uncertainty in Estuarine and Near Coastal Restoration Projects," *Journal of Coastal Research* 40 (2005): 94–108.

17. Morone and Woodhouse, *The Demise of Nuclear Energy?*, 133–135.

18. Dotson and Bouchey, "Democracy and the Nuclear Stalemate."

19. Hansman, "Re-Engineering the Colorado River."

20. Lawrence Susskind, Alejandro E. Camacho, and Todd Schenk, "A Critical Assessment of Collaborative Adaptive Management in Practice," *Journal of Applied Ecology* 49, no. 1 (2012): 47–51; and Lawrence Susskind, Alejandro E. Camacho, and Todd Schenk, "Collaborative Planning and Adaptive Management in Glen Canyon: A Cautionary Tale," *Columbia Journal of Environmental Law* 51, no. 1 (2010): 1–55.

21. Patrick Barkham, "'You Can't Control What Beavers Do or How They Do It!' Could Rewilding Help England Fight Droughts," *Guardian*, August 24, 2022, https://www.theguardian.com/environment/2022/aug/24/you-cant-control-what-beavers-do-or-how-they-do-it-could-rewilding-help-england-fight-droughts.

22. Deborah M. Coz and Juliette C. Young, "Conflicts Over Wildlife Conservation: Learning from the Reintroduction of Beavers in Scotland," *People and Nature* 2, no. 2 (2020): 406–419.

23. Ilona Amos, "Fears Raised as 87 Protected Beavers Legally Shot in Tayside," *Scotsman*, May 28, 2020, https://www.scotsman.com/news/environment/fears-raised-as-87-protected-beavers-legally-shot-in-tayside-2867381.

24. For the sources of wind turbine resistance, see Joseph Rand and Ben Hoen, "Thirty Years of North American Wind Energy Acceptance: What Have We Learned?," *Energy Research & Social Science* 29 (2017): 135–148; Claude Pied, "Negotiating the Northwoods: Anti-Establishment Rural Politics in the Northeastern United States," *Journal of Rural Studies* 82 (2021): 293–302; and Salam Elmallah and Joseph Rand, "'After the Leases Are Signed, It's a Done Deal': Exploring Procedural Injustices for Utility-Scale Wind Energy Planning in the United States," *Energy Research & Social Science* 89 (2022): 102549.

25. Barkham, "'You Can't Control What Beavers Do.'"

26. Zach St. George, *The Journey of Trees: A Story about Forests, People, and the Future* (New York: W. W. Norton & Company, 2020), 1–3.

27. St. George, *The Journey of Trees*, 9. See also Marris, *Rambunctious Garden*, 81–83.

28. Miguel Fernández, Healy H. Hamilton, and Lara M. Kueppers, "Back to the Future: Using Historical Climate Variation to Project Near-Term Shifts in Habitat Suitable for Coast Redwood," *Global Change Biology* 21, no. 11 (2015): 4141–4152.

29. Michelle Nijhuis, "Taking Wilderness in Hand: Rescuing Species," *Orion*, accessed October 16, 2023, https://orionmagazine.org/article/rescuing-species/.

30. Thomas, *Inheritors of the Earth*, 70–76.

31. St. George, *The Journey of Trees*, 195.

32. Anthony Ricciardi and Daniel Simberloff, "Assisted Colonization Is Not a Viable Conservation Strategy," *Trends in Ecology & Evolution* 24, no. 5 (2009): 248–253.

33. Brian J. Palik, Peter W. Clark, Anthony W. D'Amato, Chris Swanston, and Linda Nagel, "Operationalizing Forest-Assisted Migration in the Context of Climate Change Adaptation: Examples from the Eastern USA," *Ecosphere* 13, no. 10 (2022): e4260.

34. Nathalie Butt, Alienor L. M. Chauvenet, Vanessa M. Adams, Maria Beger, Rachael V. Gallagher, Danielle F. Shanahan, et al., "Importance of Species Translocations Under Rapid Climate Change," *Conservation Biology* 35, no. 3 (2021): 775–783.

35. St. George, *The Journey of Trees*, 110–112; and Evelyn Brister, J. Britt Holbrook, and Megan J. Palmer, "Conservation Science and the Ethos of Restraint," *Conservation Science and Practice* 3, no. 4 (2021): e381.

36. Masksim Lavrik, "Current Conservation Regimes and the Road to Laws on Assisted Migration," *Environmental Management* 69 (2022): 1186–1201.

37. Greg Breining, "New Forest for a New Climate," *Minnesota Conservation Volunteer*, March–April 2022, https://www.dnr.state.mn.us/mcvmagazine/issues/2022/mar-apr/forest.html.

38. "Coral Spawning" by Pei Yan, *flickr*, April 14, 2018, https://flic.kr/p/26aVerX, used under Creative Commons Attribution-NoDerivs 2.0 Generic license, https://creativecommons.org/licenses/by-nd/2.0/.

39. Line K. Bay, Andrew Heyward, and Andrew Negri, "Explainer: Mass Coral Spawning, A Wonder of the Natural World," *The Conversation*, November 15, 2017, https://theconversation.com/explainer-mass-coral-spawning-a-wonder-of-the-natural-world-87253; and Kolbert, *The Sixth Extinction*, 145–147.

40. Bay et al., "Explainer: Mass Coral Spawning"; and Australian Institute of Marine Science, "Coral Spawning Excitement in the National Sea Simulator," YouTube, 1:06, December 3, 2018, https://www.youtube.com/watch?v=Rtmz8yXmXvo.

41. Warren Cornwall, "Researchers Embrace a Radical Idea: Engineering Coral to Cope with Climate Change," *Science*, March 21, 2019, https://www.science.org/content/article/researchers-embrace-radical-idea-engineering-coral-cope-climate-change.

42. Cornwall, "Researchers Embrace a Radical Idea."

43. Liam Lachs et al., "Emergent Increase in Coral Thermal Tolerance Reduces Mass Bleaching Under Climate Change," *Nature Communications* 14 (2023): 4939.

44. Cornwall, "Researchers Embrace a Radical Idea."

45. Gabriel Popkin, "Can Genetic Engineering Bring Back the American Chestnut?," *New York Times*, April 30, 2020, https://www.nytimes.com/2020/04/30/magazine/american-chestnut.html.

46. Mistakes have been made. It was recently discovered that most of the trees in experimental plots were the wrong engineered variant—version D54 rather than D58. Other researchers believe that a different genetically engineered line or even plain-old selectively breed chestnut trees will be more promising. Lela Nargi, "A Lab Error Just Upended Years of Work on the GMO American Chestnut," *Ambrook Research*, January 20, 2024, https://ambrook.com/research/sustainability/GMO-chestnut-controversy.

47. Gabriel Popkin, "To Save Iconic American Chestnut, Researchers Plan Introduction of Genetically Engineered Tree into the Wild," *Science*, August 29, 2018, https://www.science.org/content/article/save-iconic-american-chestnut-researchers-plan-introduction-genetically-engineered-tree; Elizabeth Kolbert, "CRISPR and the Splice to Survive," *New Yorker*, January 11, 2021, https://www.newyorker.com/magazine/2021/01/18/crispr-and-the-splice-to-survive; and American Chestnut Foundation, "Darling 58," accessed March 27, 2024, https://tacf.org/darling-58/.

48. Popkin, "Can Genetic Engineering Bring Back the American Chestnut?"

49. Kavin Senapathy, "I Was Lured into Monsanto's GMO Crusade. Here's What I Learned," *Undark*, June 27, 2019, https://undark.org/2019/06/27/monsanto-gmo-crusade/.

50. Carey Gillam, "Appeals Court Rejects Bayer Bid to Overturn Roundup Trial Loss; Cites Monsanto 'Reckless Disregard' for Consumer Safety," *U.S. Right to Know*, August 9, 2021,

https://usrtk.org/monsanto-roundup-trial-tracker/another-setback-for-bayer-appeals-court-rejects-company-bid-to-overturn-loss-in-pilliod-trial/.

51. Lessley Anderson, "Why Does Everyone Hate Monsanto?," *Modern Farmer*, March 4, 2014, https://modernfarmer.com/2014/03/monsantos-good-bad-pr-problem/; Jessica C. Barnes and Jason A. Delborne, "The Politics of Genetic Technoscience for Conservation: The Case of Blight-Resistant American Chestnut," *Environment and Planning E: Nature and Space* 5, no. 3 (2022): 1518–1540; and Amy Harmon, "A Lonely Quest for Facts on Genetically Modified Crops," *New York Times*, January 4, 2014, http://www.nytimes.com/2014/01/05/us/on-hawaii-a-lonely-quest-for-facts-about-gmos.html?_r=0.

52. Barnes and Delborne, "The Politics of Genetic Technoscience"; S. Kathleen Barnhill-Dilling and Jason A. Delborne, "Who Intentions? What Consequences? Interrogating 'Intended Consequences' for Conservation with Environmental Biotechnology," *Conservation Science and Practice* 3 (2021): e406.

53. Jason A. Delborne, Adam E. Kokotovich, and Jeantine E. Lunshof, "Social License and Synthetic Biology: The Trouble with Mining Terms," *Journal of Responsible Innovation* 7, no. 3 (2020): 280–197.

54. Cornwall, "Researchers Embrace a Radical Idea."

55. James Spurgeon, "The Value of Our Vanishing Reefs," *Land Journal* March–April (2019): 20–21.

56. Robert Davis, Bill Bateman, Damian Lettoof, Maggie J. Watson, and Michael Lohr, "Mouse Plague: Bromadiolone Will Obliterate Mice, But It'll Poison Eagles, Snakes and Owls, Too," *The Conversation*, May 20, 2021, https://theconversation.com/mouse-plague-bromadiolone-will-obliterate-mice-but-itll-poison-eagles-snakes-and-owls-too-160995.

57. Ian Mannix, "World's First Trial to Eradicate Mice Through Gene Modification," *Cosmos*, November 10, 2022, https://cosmosmagazine.com/nature/gene-trial-to-eradicate-mice/; Ed Yong, "New Zealand's War on Rats Could Change the World," *Atlantic*, November 16, 2017, https://www.theatlantic.com/science/archive/2017/11/new-zealand-predator-free-2050-rats-gene-drive-ruh-roh/546011/; and Kate Guthrie, "Study Looks at Kiore when Rival Rat Species Absent," *PredatorFreeNZ*, October 8, 2020, https://predatorfreenz.org/research/kiore-slipper-island/.

58. "Gene drive" by Mariuswalter, *Wikimedia*, September 25, 2017, used under Creative Commons-Share Alike 4.0 International license, https://creativecommons.org/licenses/by-sa/4.0/deed.en. Modifications by author: Conversion to grayscale, addition of descriptive text, and inclusion of visualization of genetic modification using a public domain double helix graphic. Source regarding gene drive function: Megan Scudellari, "Self-Destructing Mosquitoes and Sterilized Rodents: The Promise of Gene Drives," *Nature* 571, no. 7764 (2019): 160–162.

59. Ethan Bier, "Gene Drives Gaining Speed," *Nature Reviews Genetics* 23, no. 1 (2022): 5–22; and Shayla Love, "This Gene Technology Could Change the World. Its Maker Isn't Sure

It Should," *Vice*, November 18, 2019, https://www.vice.com/en/article/kz49ge/this-gene-technology-could-change-the-world-its-maker-isnt-sure-it-should-v26n4.

60. Luke Gierus, Aysegul Birand, Mark D. Bunting, Gelshan I. Godahewa, Sandra G. Piltz, Kevin P. Oh, et al., "Leveraging a Natural Murine Meiotic Drive to Suppress Invasive Populations," *Proceedings of the National Academy of Sciences* 119, no. 46 (2022): e2213308119.

61. Rowan Jacobsen, "Deleting a Species," *Pacific Standard*, September 7, 2018, https://psmag.com/magazine/deleting-a-species-genetically-engineering-an-extinction.

62. Jacobsen, "Deleting a Species"; and Yong, "New Zealand's War on Rats."

63. Kevin M. Esvelt and Neil J. Gemmell, "Conservation Demands Safe Gene Drive," *PLoS Biology* 15, no. 11 (2017): e2003850; and MIT Media Lab, "Daisy-Chain Gene Drives," YouTube, April 3, 2019, 5:14, https://www.youtube.com/watch?v=n71fcuFtbgQ.

64. As with nuclear weapons, such a contradiction makes efforts to secure the safety of gene drives inherently fraught. See Scott D. Sagan, *The Limits of Safety: Organizations, Accidents, and Nuclear Weapons* (Princeton, NJ: Princeton University Press, 1993).

65. Bier, "Gene Drives Gaining Speed."

66. Lizette Alvarez, "A Mosquito Solution (More Mosquitos) Raises Heat in Florida Keys," February 19, 2015, https://www.nytimes.com/2015/02/20/us/battle-rises-in-florida-keys-over-fighting-mosquitoes-with-mosquitoes.html; and Sarah Kuta, "First U.S. Open-Air Test of Genetically Modified Mosquitos Deemed a Success," *Smithsonian Magazine*, April 21, 2022, https://www.smithsonianmag.com/smart-news/first-us-open-air-test-of-genetically-modified-mosquitoes-deemed-a-success-180979960/.

67. Charles Perrow, *Normal Accidents* (Princeton, NJ: Princeton University Press, 1984); and Charles Perrow, "Fukushima and the Inevitability of Accidents," *Bulletin of the Atomic Scientists* 67, no. 6 (2011): 44–52.

68. Dotson and Bouchey, "Democracy and the Nuclear Stalemate"; and Trevor A. Kletz, "What You Don't Have Can't Leak," *Chemistry and Industry* 6 (1978): 287–302.

69. Kolbert, "CRISPR"; and Kolbert, *Under a White Sky*, 124–125.

70. Henry Bodkin, "Cats Could Have the Hunting Instinct 'Edited' Out of Their Genes," *The Telegraph*, June 12, 2016, https://www.telegraph.co.uk/science/2016/06/12/cats-could-have-the-hunting-instinct-edited-out-of-their-genes/.

71. Abi Gazzard, Connor Panter, and Rosalind Kennerley, "In Defence of Rodents—Why Healthy Ecosystems Need Them," *The Conversation*, January 3, 2023, https://theconversation.com/in-defence-of-rodents-why-healthy-ecosystems-need-them-196406.

72. Mannix, "World's First Trial to Eradicate Mice."

73. "Lyme Disease: Data and Surveillance," *CDC*, August 29, 2022, https://www.cdc.gov/lyme/datasurveillance/index.html#:~:text=Standard%20national%20surveillance%20is%20only,year%20in%20the%20United%20States.

74. Ed Yong, "One Man's Plan to Make Sure Gene Editing Doesn't Go Haywire," *Atlantic*, July 11, 2017, https://www.theatlantic.com/science/archive/2017/07/a-scientists-plan-to-protect-the-world-by-changing-how-science-is-done/532962/. For more on the risks of the effort, see Allison A. Snow, "Genetically Engineering Wild Mice to Combat Lyme Disease: An Ecological Perspective," *BioScience* 69, no. 9 (2019): 746–756.

75. David Abel, "In an Effort to Curb Lyme Disease, Scientists Hope to Release Thousands of Genetically Altered Mice on Nantucket," *Boston Globe*, April 21, 2022, https://www.bostonglobe.com/2022/04/21/science/an-effort-curb-lyme-disease-scientists-hope-release-thousands-genetically-altered-mice-nantucket/.

76. "Thylacine at Beaumaris Zoo, 1936" by Ben Sheppard, *GetArchive*, https://garystockbridge617.getarchive.net/media/thylacine-at-beaumaris-zoo-1936-ns4371-1-1063-354fb6.

77. John Pickrell, "Tasmanian Tiger Genome May Be Last Step Toward De-Extinction," *National Geographic*, December 11, 2017, https://www.nationalgeographic.com/science/article/thylacine-genome-extinct-tasmanian-tiger-cloning-science.

78. Scott Neuman, "Lack of Genetic Diversity May Have Doomed Tasmanian Tiger, Scientists Say," *NPR*, December 12, 2017, https://www.npr.org/sections/thetwo-way/2017/12/12/570096984/lack-of-genetic-diversity-may-have-doomed-tasmanian-tiger-scientists-say.

79. John Timmer, "This De-Extinction Company Wants to Resurrect the Thylacine," *Wired*, August 17, 2022, https://www.wired.com/story/thylacine-mammoth-de-extinction-colossal/.

80. Timmer, "This De-Extinction Company."

81. Charles Y. Feigin, Axel H. Newton, Liliya Doronina, Jürgen Schmitz, Christy A. Hipsley, Kieren J. Mitchell, et al., "Genome of the Tasmanian Tiger Provides Insights into the Evolution and Demography of an Extinct Marsupial Carnivore," *Nature Ecology & Evolution* 2, no. 1 (2018): 182–192; and Neuman, "Lack of Genetic Diversity."

82. Dolly Jørgensen, "Reintroduction and De-Extinction," *BioScience* 63, no. 9 (2012): 719–720.

83. Sam Walters, "A New Approach Allows Scientists to Extract DNA from Pickled Specimens," *Discover*, July 5, 2022, https://www.discovermagazine.com/the-sciences/a-new-approach-for-extracting-dna-from-pickled-specimens; and Justin M. Berstein and Sara Ruane, "Maximizing Molecular Data from Low-Quality Fluid-Preserved Specimens in Natural History Collections," *Frontiers in Ecology and Evolution* 10 (2022), https://doi.org/10.3389/fevo.2022.893088.

84. Timmer, "This De-Extinction Company."

85. Jackson Ryan, "Using CRISPR to Resurrect the Dead," *CNET*, June 19, 2029, https://www.cnet.com/science/features/using-crispr-to-resurrect-the-dead/.

86. Ben Jacob Novak, "De-Extinction," *Genes* 9, no. 11 (2018): 548.

87. Sara Ord, "The Thylacine: An Exemplary Candidate for De-Extinction," *Colossal Laboratories & Biosciences*, accessed October 27, 2023, https://colossal.com/thylacine-an-exemplary-candidate-for-de-extinction/.

88. “Saving Species: Arabian Oryx,” *Phoenix Zoo*, accessed October 27, 2023, https://www.phoenixzoo.org/blog/zoo-news/saving-species-arabian-oryx/.

89. Ronald Sandler, “De-Extinction and Conservation Genetics in the Anthropocene,” *Hastings Center Report* 47 (2017): S46.

90. Novak, “De-Extinction.”

91. Such bodies are called democratic mini-publics. See Graham Smith and Maija Setälä, “Mini-Publics and Deliberative Democracy,” in *The Oxford Handbook of Deliberative Democracy*, ed. Andre Bächtiger et al. (New York: Oxford University Press, 2018), 300–314; and Victoria Herridge, “Before Making a Mammoth, Ask the Public,” *Nature* 598, no. 387 (2021), https://doi.org/10.1038/d41586-021-02844-5.

92. Political scientist Charles Lindblom associated the view that democracy was “government by discussion” with the ideal of a technocratic or “scientific society.” This is because it assumes that some group of people or deliberative mechanism “could find harmony in the universe, each person potentially no longer in conflict with others, the intellect consequently capable of discovering ‘correct’ solutions.” As you can probably gather, I agree with Lindblom in viewing the more productive pathway as lying in arranging society so as to better learn through experimental action and the mutual adjustments made between differently biased partisans. If we do find “correct” solutions, it will be via trial and error, not within the minds of technocrats or emerging out of deliberative bodies. Lindblom, *Inquiry and Change*, 228.

93. “Research,” *Thylacine Integrated Genomic Restoration Research Lab*, accessed October 27, 2023, https://tigrrlab.science.unimelb.edu.au/research/.

94. Michael Greshko, “Mammoth–Elephant Hybrids Could Be Created Within the Decade. Should They Be?,” *National Geographic*, September 13, 2021, https://www.nationalgeographic.com/science/article/mammoth-elephant-hybrids-could-be-created-within-the-decade-should-they-be.

95. Brister et al., “Conservation Science and the Ethos of Restraint.” The matter isn't helped by the fact that the “best available science” requirement of laws such as the Endangered Species Act has been interpreted by the courts to exclude approaches such as adaptive management. A precautionary principle-based framework is obviously built on nonscientific values, but it enjoys a lack of scrutiny due to being the cultural status quo. On the other hand, agencies have been criticized for using adaptive management as an excuse to delay difficult decisions. See Shana Lee Hirsch and Jerrold Long, “Non-Epistemic Values in Adaptive Management,” *Environmental Values* 27, no. 5 (2018): 467–488.

96. Aaron Wildavsky, *Searching for Safety* (New Brunswick, NJ: Transaction Books, 1988).

97. Cornwall, “Researchers Embrace a Radical Idea.”

98. Claire Hirashiki, Peter Kareiva and Michelle Marvier, “Concern Over Hybridization Risks Should Not Preclude Conservation Interventions,” *Conservation Science and Practice* 4, no. 4 (2021): e424.

CHAPTER 8

1. Stefanie A. Crabtree, Douglas W. Bird, and Rebecca Bliege Bird, "Subsistence Transitions and the Simplification of Ecological Networks in the Western Desert of Australia," *Human Ecology* 47 (2019): 165–177.
2. Crabtree et al., "Subsistence Transitions and the Simplification."
3. Crabtree et al., "Subsistence Transitions and the Simplification."
4. In Bruno Latour's final book, *Down to Earth*, he refers to something called the "Terrestrial," which seems similar to what I'm getting at here. He contrasts the terrestrial view with an expansively global and abstract perspective and with narrow localism. It is a situation in which we recognize nature as an actor, one who reacts to our presence and whose reactions we must now respond to. In my opinion, Latour relies too much on the idea that our era is uniquely terrestrial. Some might say that we've always been so, even if a French philosopher hadn't yet alerted us to that reality, for I see the Martu as doing exactly what Latour describes: reacting to nature's reactions. In any case, his criticism of the polarizing and unhelpful binary of ecology and economics leads me to believe that he would give support, albeit with stereotypically Gallic tepidness, to the vision that I've laid out with this example. See Bruno Latour, *Down to Earth: Politics of the New Climatic Regime* (New York: Polity Press, 2018).
5. Convention on Biological Diversity, "Decision Adopted by the Conference of the Parties."
6. Kolbert, *Under a White Sky*, 137.
7. Kolbert, *Under a White Sky*, 200.
8. Vonnegut, *Galápagos*, 226.
9. Ted Nordhaus, "Why We Can't Leave Nature Alone," *New Atlantis*, Spring 2021, https://www.thenewatlantis.com/publications/why-we-cant-leave-nature-alone.
10. Nordhaus, "Why We Can't Leave Nature Alone."
11. Environmental Protection Agency, "Basic Information About Sewage Sludge and Biosolids," January 18, 2023, https://www.epa.gov/biosolids/basic-information-about-sewage-sludge-and-biosolids.
12. Edward Fitzpatrick, "Rhode Island Was One of First States to Build Sewers and Treatment Plants," *Politifact*, November 28, 2010, https://www.politifact.com/factchecks/2010/nov/28/jamie-samons/rhode-island-was-one-first-states-build-sewers-and/; and Lina Zeldovich, "A History of Human Waste as Fertilizer," *JSTOR Daily*, November 18, 2019, https://daily.jstor.org/a-history-of-human-waste-as-fertilizer/.
13. Nicholas Maxwell, *From Knowledge to Wisdom: A Revolution for Science and the Humanities* (London: Pentire Press, 2007).
14. Maxwell, *From Knowledge to Wisdom*, 60–77.
15. Maxwell, *From Knowledge to Wisdom*, 95.

16. See Esther Turnhout, Claire Waterton, Katja Neves, and Marleen Buizer, "Rethinking Biodiversity: From Goods and Services to 'Living With,'" *Conservation Letters* 6, no. 3 (2012): 154–161.

17. Daniel Sarewitz, "Stop Treating Science Denial Like a Disease," *Guardian*, August 21, 2017, https://www.theguardian.com/science/political-science/2017/aug/21/stop-treating-science-denial-like-a-disease.

18. IPBES, "Foundations of the Nature Futures Framework," *Annex VI to Decision IPBES-9/1*, accessed December 1, 2023, https://www.ipbes.net/resource-file/104131.

19. One study has even found that doing so can reduce polarization-induced mistrust of scientists. Senja Post and Nis Bienyeisler, "The Honest Broker Versus the Epistocrat: Attenuating Distrust in Science by Disentangling Science from Politics," *Political Communication* 41, no. 5 (2024): 763–785.

20. Steven F. Hayward, "Is 'Conservative Environmentalist' an Oxymoron?," *American Enterprise Institute*, June 20, 2005, https://www.aei.org/research-products/speech/is-conservative-environmentalist-an-oxymoron/.

21. Hayward, "Is 'Conservative Environmentalist' an Oxymoron?"; and Christopher Sellers, "How Republicans Came to Embrace Anti-Environmentalism," *Vox*, June 7, 2017, https://www.vox.com/2017/4/22/15377964/republicans-environmentalism.

22. Rich, *Getting to Green*.

23. R. Wills Flowers, "Night and Fog: The Backlash Against the Endangered Species Act," *Wild Earth* 2, no. 3 (1992): 6–9.

24. Ted Nordhaus, Michael Shellenberger and Jenna Mukuno, "Ecomodernism and the Anthropocene," *The Breakthrough Journal*, June 15, 2015, https://thebreakthrough.org/journal/issue-5/ecomodernism-and-the-anthropocene.

25. Marco Verweij, "Is the Kyoto Protocol Merely Irrelevant, or Positively Harmful, for the Efforts to Curb Climate Change?," in *Clumsy Solutions for a Complex World: Governance Politics and Plural Perspectives*, ed. Marco Verwejj and Michael Thomas (New York: Palgrave MacMillan, 2006), 31–59.

26. Jessica Lovering and Ted Nordhaus, "Does Climate Policy Matter?," *The Breakthrough Journal*, November 28, 2016, https://thebreakthrough.org/issues/energy/does-climate-policy-matter.

27. Marco Verwejj and Michael Thomas (eds.), *Clumsy Solutions for a Complex World: Governance Politics and Plural Perspectives* (New York: Palgrave MacMillan, 2006).

28. Natalia Mesa, "Migratory Birds Needs Habitat. California Farmers Can Help," *Undark*, October 22, 2024, https://undark.org/2024/10/22/migratory-birds-need-habitat-california-farmers-can-help/; and Lisa Schulte Moore, "Prairie Strips," *Missouri Prairie Journal* 35, no. 1 (2014): 12–15.

29. National Renewable Energy Lab, "Agrivoltaics," accessed December 1, 2023, https://www.nrel.gov/solar/market-research-analysis/agrivoltaics.html; and Sarah DeWeerdt, "In the

Gentle Shade of Solar Panels, Desert Restoration Gets an Unexpected Leg Up," *Anthropocene Magazine*, April 25, 2023, https://www.anthropocenemagazine.org/2023/04/in-the-gentle-shade-of-solar-panels-desert-restoration-gets-an-unexpected-leg-up/.

30. Allaire Diamond, "Building Beaver Dam Analogs to Restore Watersheds," *Northern Woodlands*, September 1, 2023, https://northernwoodlands.org/articles/article/beaver-dam-analogs; and Laura Lundquist, "Dam Pipes Save Beavers and Ditches," *Bozeman Daily Chronicle*, May 5, 2013, https://www.bozemandailychronicle.com/news/wildlife/dam-pipes-save-beavers-and-ditches/article_3a64ba42-b53e-11e2-948f-0019bb2963f4.html.

31. Alison Martin, Anke Fischer, Rob McMorran, and Melanie Smith, "Taming Rewilding—from the Ecological to the Social: How Rewilding Discourse in Scotland Has Come to Include People," *Land Use Policy* 111 (2021): 105677.

32. One inspirational case is the use of semi-domesticated elephants in Burma. In Jacob Shell's book, *Giants of the Monsoon Forest*, he describes a system in which Asian elephants are captured and put to work in logging teak or transporting goods, when heavy monsoon waters wash out even well-built roads and bridges. The elephants aren't exactly captives. They are released back to local forests to forage, sleep, and mate with wild elephants when not working and "fetched" again in the morning. Many elephants seem to bond with their riders or at least adapt to their new occupations. Although morally fraught, the practice nevertheless takes pressure off wild populations and could provide a foundation for a conservation alternative to the NGO-driven insistence on pristine wilderness. See Jacob Shell, *Giants of the Monsoon Forest: Living and Working with Elephants* (New York: W. W. Norton & Company, 2019).

33. Hannah Ritchie, "Wild Mammals Are Making a Comeback in Europe Thanks to Conservation Efforts," *Our World in Data*, September 27, 2022, https://ourworldindata.org/europe-mammal-comeback; Johnny Wood, "The World Has More Trees Than It Did 35 Years Ago—But There's a Huge Catch," *World Economic Forum*, August 30, 2018, https://www.weforum.org/agenda/2018/08/planet-earth-has-more-trees-than-it-did-35-years-ago/; and Leslie Nemo, "From 440 to 25,000: One Humpback Whale Population's Amazing Recovery," *Discover*, November 18, 2019, https://www.discovermagazine.com/planet-earth/from-440-to-25-000-one-humpback-whale-populations-amazing-recovery.

34. Michael Oakeshott, *Rationalism in Politics and Other Essays* (London: Methuen & Co., 1962), 168–196.

35. Will Steffen, "How Defining Planetary Boundaries Can Reform Our Approach to Growth," *Resilience*, May 25, 2011, https://www.resilience.org/stories/2011-05-25/how-defining-planetary-boundaries-can-transform-our-approach-growth/; and Emma Marris, "How to Save Endangered Species Without a Plan for Every Bee," *The Atlantic*, May 25, 2025, https://www.theatlantic.com/science/archive/2025/05/endangered-species-harm-ecosystems/682883/.

36. Oakeshott, *Rationalism in Politics*, 186.

37. Graham Saul, *Environmentalists, What Are We Fighting For?* (Toronto, ON: Metcalf Foundation, 2018), 6.

38. Daniel Botkin, *The Moon in the Nautilus Shell: Discordant Harmonies Reconsidered* (New York: Oxford University Press, 2012), 12.

39. Zachary Nickell, Sofia Varriano, Eric Plemmons, and Matthew D. Moran, "Ecosystem Engineering by Bison (*Bison bison*) Wallowing Increases Arthropod Community Heterogeneity in Space and Time," *Ecosphere* 9, no. 9 (2018): e02436.

40. As one group of behavioral scientists put it, "Individual-level solutions may support system-level change if they are internalized as part of one's social identity." People beginning to take action, even trivially helpful steps toward creating a wilder world and having a positive experience of doing so, is likely more persuasive and transformative than them reading an editorial from a scientist or a debunking article. See Lina Koppel, Claire E. Robertson, Kimberly C. Doell, Ali M. Javeed, Jesper Rasmussen, Steve Rathje, et al., "Individual-Level Solutions May Support System-Level Change if They Are Internalized as Part of One's Social Identity," *Behavioral and Brain Science* 46 (2023): e165.

41. Lindblom, *Inquiry and Change*, 244.

42. Richard Rorty, *Achieving Our Country* (Cambridge, MA: Harvard University Press, 1998), 52.

Index

Publisher contact:
The MIT Press
Massachusetts Institute of Technology
77 Massachusetts Avenue, Cambridge, MA 02139
mitpress.mit.edu

EU Authorised Representative:
Easy Access System Europe, Mustamäe tee 50,
10621 Tallinn, Estonia
gpsr.requests@easproject.com

Printed by Integrated Books International,
United States of America